The Environment of Oil

STUDIES IN INDUSTRIAL ORGANIZATION

Volume 17

THE ENVIRONMENT OF OIL

edited by
Richard J. Gilbert

KLUWER ACADEMIC PUBLISHERS
Boston / Dordrecht / London

Distributors for North America:
Kluwer Academic Publishers
101 Philip Drive
Assinippi Park
Norwell, Massachusetts 02061 USA

Distributors for all other countries:
Kluwer Academic Publishers Group
Distribution Centre
Post Office Box 322
3300 AH Dordrecht, THE NETHERLANDS

Library of Congress Cataloging-in-Publication Data
The Environment of oil/edited by Richard J. Gilbert.
      p.   cm.
  Includes index.
  ISBN 0-7923-9287-6 (alk. paper)
  1. Petroleum industry and trade--Environmental aspects--United
States. 2. Petroleum industry and trade--Economic aspects--United
States. 3. Energy policy--United States. I. Gilbert, Richard J.,
1945-
TD195.P4E52 1993
333.79'6814--dc20   92-21475
                                           CIP

Printed on acid-free paper.

Printed in the United States of America

TABLE OF CONTENTS

ACKNOWLEDGMENTS

This book is the end result of a research program supported by the University of California Energy Institute, intended to examine aspects of the outlook for oil supply and demand and the conflicts between environmental protection and economic growth. We are indebted to the support made available by the University of California and to many of our colleagues who made this effort possible. Mike Lederer, Deputy Director of the Energy Institute, played a major role in the organization of the project and in editing the final product. Mike's deep understanding of energy policy issues and his attention to detail helped focus the project on important questions and bring the results together in this volume. Mike made numerous contributions to our research efforts and worked with our able typesetter and copy editor, Lindsay Mugglestone, to produce the final product. The index was produced by Paul Kish.

Many of our friends and colleagues provided useful suggestions during the course of this project. We owe particular thanks to Carl Blumstein, Severin Borenstein, and Tracy Lewis, who participated in many of the project meetings and provided valuable input to research directions. Others who helped us along the way include Anthony Brown, Lee Friedman, Connie Helfat, Philip Judson, Ed Kahn, Bart McGuire, Roger Noll, John Quigley, Jim Sweeney, David Teece, and Len Waverman. We owe a special debt to our Energy Institute administrative staff, Carol Kozlowski and Linda Dayce, who helped organize meetings, coordinate communications, prepare working drafts, and generally make things happen. We could not have done it without them.

Finally, I have to thank my colleagues whose names appear in this volume. Although we did not always agree on every issue, they persisted in this endeavor despite our differences and brought the project to a successful conclusion. Most of all, they contributed to an exciting and stimulating environment for the analysis of these important issues.

R. J. Gilbert

1 INTRODUCTION TO THE ENVIRONMENT OF OIL

Richard J. Gilbert

The problems of relying on oil as the propellant of the U.S. economy occupy a central position in national energy policy debates. Oil carries with it reliance on unstable sources of supply and the potential for environmental catastrophe. It is a finite resource, and although crude oil was no more expensive (in real terms) in 1991 than it was in 1974, few expect that the recent experience of stable oil prices will persist into the next century. Oil, many would argue, has succeeded all too well as the world's premier energy source. Industrialized nations have evolved economies that are dependent on large quantities of affordable petroleum products. Byproducts of our petroleum dependence are urban atmospheres that are fouled by noxious automobile emissions and cities that are strangled by traffic congestion.

The purpose of this book is to look closely at some of the controversial issues that surround the use of oil as an energy source and to see whether there is anything new that can be said to illuminate public policy in this vital area. It is not an attempt to write a complete survey of the economic performance and environmental impacts of our petroleum economy. Although the coverage of this book is incomplete, our intent is to address a wide range of issues that affect our society and its dependence on oil-based fuels. By "the environment of oil" we mean a wide spectrum of social structures and impacts that are a part of the petroleum economy. In addition to the more familiar topics such as

Richard J. Gilbert is with the Department of Economics, University of California, Berkeley, and Director of the California Energy Institute.

balancing energy needs against the environmental cost of oil spills, we also include such topics as how well the industry succeeds in delivering petroleum products to consumers at prices that approximate costs.

Our central conclusion is that all is not doom and gloom in the outlook for world oil supply and demand. The reasons for an optimistic short-term outlook are large world crude reserves and a decline in the rate of increase in the demand for petroleum fuels in transportation (in the U.S. and possibly other industrialized countries), the single most important component of petroleum use. Over the longer term, price increases are likely to lead to more investment in petroleum exploration, which will help sustain the industry record of reserve additions, although production declines in the continental United States are inevitable in the near future. Higher prices also will lead to more efficient use of petroleum fuels. These factors, along with sizable gaps between the current delivered price of oil and the delivered cost of realistic alternative fuels, imply that the petroleum economy will continue to be the engine of economic growth for many decades to come.

This book is the end result of a study project funded by the Energy Research Institute of the University of California. In keeping with its origins, the book pays particular attention to petroleum issues as they impact consumers in the state of California. We recognize that oil is produced and consumed in a world market and that California is not an island in the world market; however, California has been a leader in many crucial areas of energy policy. California is the nation's fourth largest oil producer. Air quality in California's major urban areas is threatened by enormous amounts of vehicular traffic and by geographic and climatic conditions that concentrate urban smog. California has responded with the toughest air quality standards in the nation and with regulations to promote the use of alternatives to petroleum fuels.

The dimension of the problem of securing adequate supplies of petroleum or petroleum substitutes depends, of course, on the magnitudes of the demands for energy-intensive products and services. Although petroleum has many uses, it is generally recognized that the crucial component of petroleum demand is its use as a transportation fuel. This is the subject of Chapter 2, by David Brownstone and Charles Lave. In the United States, oil accounts for over 95% of the energy used in transportation and transportation accounts for over 60% of the demand for petroleum fuels. Consumption of motor gasoline in the U.S. grew by 4.7% per year from 1950 through 1973. This rate of growth implies a doubling of gasoline consumption every 15 years. With gasoline accounting for almost half of all petroleum consumption and with no economically viable near-term alternative to conventional gasoline-powered vehicles for most transportation needs, a doubling of gasoline demand every 15 years would place an intolerable burden on the ability of world oil supply to keep pace with demand. Although the demand for gasoline has slowed in recent years, there is much concern that demand will return to the rapid growth that has character-ized much of the post-War period. Fortunately, Brownstone and Lave uncover several reasons why this is an unlikely scenario.

Improvements in Corporate Average Fuel Economy (CAFE) requirements have occupied much of the transportation policy debate in the U.S. Improve-ments in fleet average fuel efficiency over the past two decades have had a significant impact on gasoline demand. Since 1973, average automobile pas-senger car efficiency has increased by more than 50%. Further increases in fuel efficiency are likely as a result of market-driven technological gains and legis-lated improvements. However, Brownstone and Lave argue that higher CAFE standards are unlikely to be the dominant factor in the rate of growth of gas-oline demand over the next two decades. CAFE standards affect only new vehicles, and slow turnover of the U.S. automobile fleet results in only a mod-erate effect on the average fuel economy of the entire fleet. Brownstone and Lave cite projections that show that average fuel economy in California in 2010 would be about 17.3 MPG without a 40 MPG CAFE requirement and 19.2 MPG with the proposed requirement.

Brownstone and Lave also argue that increased emphasis on mass transit would have only a small effect on the demand for transportation fuels. In recent years, the difference in energy consumed per passenger-mile has narrowed for different transportation modes. Commuter buses and trains have higher load factors than commuter autos, which makes than more energy-efficient. How-ever, the fraction of transportation fuel demand that would be affected by mass transit and the size of the efficiency gap between mass transit and autos are not large enough to have major impacts on total energy use. Thus public policy toward mass transit should emphasize its contribution to reducing urban con-gestion and improving transportation access, rather than its effects on the de-mand for petroleum.

Brownstone and Lave expect the demand for gasoline to slow dramatically in the next two decades, but for reasons that have little to do with CAFE standards and mass transit. From 1950 until 1990, the number of vehicles driven has increased dramatically. Factors such as increased entry of women in the labor force have contributed to a more automobile-dependent society. But there is a limit to how long this trend can continue. Vehicle ownership in the U.S. is nearing the saturation point of one vehicle per licensed driver. Beyond this threshold, additional vehicle ownership cannot contribute to increased fuel consumption (and may lower consumption to the extent that new vehicles replace older and less efficient models). Brownstone and Lave are less sanguine about the prospects for a slowing in the demand for aviation fuel, but they note the significant efficiency improvements in aviation. The fuel efficiency of air transportation is approaching the same level as for automobiles, and further improvements would make air travel a more efficient transportation mode.

Although the transportation demand for energy will continue to be a primary determinant of the need for petroleum-based fuels or fuels based on petroleum substitutes, the magnitude of this demand will not be as large as one would expect by simply extrapolating the experience of the post-War period. This is the good news. The bad news is that Brownstone and Lave see little evidence of change in the U.S. away from reliance on the automobile. Vehicle miles traveled per person has remained about constant for many years and may have increased recently. There is no obvious change in commute patterns. Mass transit systems offer the potential to ease urban congestion, but Brownstone and Lave see little potential for mass transit to ease significantly the demand for transportation fuels.

The next two chapters in this book survey some of the supply options that are available to power our petroleum-based economy. The most immediate supply option is petroleum itself. In Chapter 3 Walter Mead critically examines what is publicly known about the potential for new crude oil supply in the U.S. and other parts of the world. Oil is a finite resource, but world oil reserves have shown almost a continuous increase throughout the history of the industry. These reserves represent only estimates of economic production from known fields and do not allow a projection of the total recoverable resource base. Nonetheless, Mead is optimistic about the ability of the industry to meet demand over the next two decades at (real) prices comparable to historical levels. Existing reserves are sufficient for 80 years of supply at current consumption levels, and new discoveries have exceeded increases in demand.

While world oil reserves are adequate for the near term, it is all too obvious that reserves are concentrated in unstable regions and are likely to become more concentrated. The United States is the second largest producer of oil in the world, but it is only the seventh largest in remaining known reserves. With most of the producing areas of the U.S. well-explored, and with its low ratio of known reserves to production, the United States' rank as a major crude oil producer is destined to fall. Oil self-sufficiency in the U.S. is expected to decline from 58% in 1990 to only 36% by 2010. The most likely prospects for a substantial increase in U.S. petroleum production capacity are in the Alaskan National Wildlife Range and the Santa Maria Basin off the coast of California. Exploration in both areas faces stiff environmental opposition.

The West Coast of the United States has enjoyed a petroleum surplus caused by restrictions on the export of oil from the Alaska North Slope (ANS). The only practical alternative to selling the oil on the West Coast of North America is to transport it, at higher cost, to the Gulf Coast and the East. This surplus is likely to disappear in the next decade, and along with it the small price advantage that consumers on the West Coast had enjoyed relative to their neighbors to the east. It is unlikely, however, that the magnitude of this effect will be more than $1 to $2 per barrel—about 2 to 5 cents per gallon of gasoline. One of the few bright spots in U.S. crude oil supply is the potential for a near-term increase in production of heavy oil, particularly from the central regions of California.

World oil supplies are adequate to fuel economic growth for many decades, and oil prices must increase significantly before a transition to replacement sources of energy becomes economically feasible. Measures such as the national synthetic fuels program undertaken by President Carter, which sought to produce petroleum fuels from coal and natural gas in commercial quantities, are not likely to be economic for many years. Mead reports that synthetic fuels demonstration plants have production costs in the range of $50 to $90 per barrel, far above estimates of the cost of imported oil for next few decades. The problem facing the U.S. is not the increasing scarcity of oil, but rather that oil supplies are becoming increasingly concentrated in politically unstable regions. It is easy to confuse the two issues of short- and long-term vulnerability. The fact that oil prices move up sharply in response to an actual or perceived disruption in supply does not mean that oil has become any less valuable as a resource over the long term. It does, however, confirm that the risk of a supply disruption is an additional cost of relying on oil as an energy source. Alternative sources of supply, such as synthetic fuels, may reduce long-term oil vulnerability, but cannot have a significant impact on the security of petroleum supply for many years to come. What the United States needs is an effective program to deal with short-term supply interruptions such as occurred during the 1991 Persian Gulf conflict. The U.S. Strategic Petroleum Reserve was originally planned to have a capacity of 1 billion barrels, but actually held about 590 million barrels in 1990. This is sufficient to replace the loss of OPEC imports for about 140 days. In addition to the existing strategic reserve, Mead considers converting the Elk Hills petroleum field to a strategic reserve, leasing offshore lands, developing substitute sources of supply, and lifting the ban against exports of Alaskan crude.

Chapter 4 moves on from the subject oil supply in the short and long term to the question of what alternatives are available as petroleum substitutes. The focus is on transportation fuels, because that is the area where the alternatives seem most elusive. Automobiles are powered by petroleum and automobiles power the demand for oil. There has been much speculation about the future of the automobile: whether our economy can continue to rely on the private auto as its main transportation source, and whether we can continue to rely on a supply of energy that is vulnerable over both the short and long term.

The discussion in this chapter should be considered in light of the con-clusions in chapters 2 and 3, which describe the future demand for transporta-tion services and attempt to estimate the cost of reliance on an insecure world petroleum supply. Daniel Sperling and Mark DeLuchi survey the different en-ergy sources that could credibly serve as a substitute for oil as a transportation energy source. Sperling and DeLuchi show how different fuels are expected to perform in production of hazardous emission, including nitrogen oxides (NO_x), sulfur oxides, particulates, reactive organics, carbon monoxide, and greenhouse gases (predominantly carbon dioxide). Sperling and DeLuchi show a wide variation of emission characteristics for different fuels and points out that many fuels do not improve over conventional gasoline in all emission categories. The other dimension that Sperling and DeLuchi consider is antici-pated cost. They do not attempt a detailed forecast of the economics of alterna-tive fuels, but rather recognize the uncertainties that are inevitable with regard to the cost and performance (such as emissions, corrosive effects, energy stor-age, etc.) of petroleum substitutes.

Sperling and DeLuchi are able to rank some petroleum alternatives by their economic viability in the near term. They show that reformulated gasolines offer the potential for significant improvements in emissions performance relative to more exotic alternatives such as methanol, yet have the advantage of being a "conventional" fuel that would not require alterations in engine design and could be produced with only modest additional cost. They also show that compressed natural gas is currently available at prices that are competitive with gasoline and yet offers improved emission characteristics. One uncertainty regarding compressed natural gas, however, is the extent of price increases for this energy source if it obtained a significant share of the transportation market.

A main conclusion of Sperling and DeLuchi's work in Chapter 4 is that there is no obvious preferred alternative to gasoline and that the most attractive alternatives will depend on the environmental costs of different pollutants. Will carbon monoxide be more important than NO_x emissions? How harmful are these emissions and how much is society willing to pay to avoid them? The answers to these questions are crucial in determining which fuels should emerge as the favorites in the future and is central to any policies that are intended to encourage the introduction of alternatives to gasoline. Sperling and DeLuchi's description of alternatives to conventional gasoline-powered vehicles, and their costs and benefits, underscores the need for a better understanding of the costs of vehicle emissions and the need to incorporate these costs into the expense of vehicle ownership.

The automobile accounts for most of the air pollution problems in urban areas. National and regional emissions standards have succeeded in reducing sharply the permissible tailpipe emissions of pollutants from recent vintage automobiles, but as Alan Krupnick argues in Chapter 5, the improvements in tailpipe emissions have not been matched by improvements in urban air quality. Krupnick examines the reasons why emission standards have not been more successful in improving air quality. Emissions standards are directed primarily to new or recent automobile vintages. They do not deal adequately with older vehicles or vehicles with mechanical problems (and/or poisoned catalysts), which can be far out of compliance with emissions standards. The emissions from only a few of these vehicles can offset improvements from large numbers of vehicles that comply with existing standards.

Another problem that Krupnick considers is the focus on tailpipe emissions to the exclusion of other potentially significant vehicular emission sources. An example is evaporative emissions from gas tanks. Emissions standards may not adequately reflect the incremental harm caused by different pollutants. There may be too much emphasis on volatile organic compounds, and not enough on NO_x. At the same time, some standards may be too stringent, and in some cases (such as for ozone) almost unreachable.

Krupnick examines current policy approaches to air pollution management. Nowhere are the costs and benefits of the automobile more evident than in the Los Angeles basin, and in Chapter 5 Krupnick reviews the effectiveness of California South Coast air pollution policies and considers whether similar results could be achieved at significantly lower costs. Krupnick argues that cost-benefit evaluations are frequently ignored in the establishment of air pollution standards and that compliance could be improved by providing incentives to meet targets, rather than by setting high targets that are often not met. Among the policy alternatives considered in this chapter are gasoline fees, emission taxes, and vehicle targeting programs, all of which can be (and should be) tailored to local environmental needs.

Chapter 6, by Margriet Caswell, takes a broader look at policy approaches to combat adverse environmental impacts from the production and transportation of petroleum resources. Caswell observes that government plays a duel role as both the protector of the environment and, in many cases, the owner of petroleum resources. This duality means that government must strike a delicate balance between the desire to generate revenues and the desire to protect the environment. A major source of pollution in the use of oil occurs in the movement of oil from the wellhead to refining centers, and in the movement of refined products to distribution points. Caswell focuses on the costs and benefits of environmental regulations in petroleum transportation. Several case studies, including the 1969 Santa Barbara oil spill, the construction of the Trans-Alaska pipeline, the Long Beach–Midland pipeline proposal, and the development of the Point Arguello oil field serve to illustrate environmental concerns and to help us understand the longer-term impacts of petroleum development on ecosystems.

Caswell notes that several regulations impact the petroleum industry in ways that tend to increase environmental hazards. For example, the ban on exports of Alaskan crude oil, which is intended to protect Americans from supply vulnerability (but does not), increases tanker traffic on the West Coast of the

U.S. and therefore increases the risk of environmental damage to coastal communities.

The final two chapters of this book deal with several parts of the supply chain involved in the delivery of petroleum products to the consumer. We include this as a part of our "environment of oil" because their performance affects the value of the nation's oil resources and the price that consumers pay for oil products, which are important components of consumer satisfaction with the oil industry.

Federal and state governments own large amounts of this nation's petroleum resources. Rights to the development of these resources are transferred through lease auctions. A long-standing public policy question is whether governments have been effective in obtaining a fair compensation for the petroleum properties that are offered for private development. Walter Mead examines this question in Chapter 7 and weighs the advantages of alternative leasing schemes. Government petroleum leases are commonly awarded to private developers through bonus bidding. The right to a petroleum property is awarded to the developer who tenders the highest up-front payment. This scheme has been criticized because it has allowed some developers to earn supra-normal profits on tracts that did not attract much bid activity, yet turned out to contain significant petroleum reserves. It has been argued that an alternative bidding systems would provide a closer correspondence between the value of petroleum properties and the payments that the government collects for these properties.

Mead argues that it is wrong to focus on the difference between payment and value for individual properties. The relevant question is whether the government receives total payments from all private developers that are close to the total value of the properties that are offered for lease. He shows that bonus bids for government petroleum properties have been, in the aggregate, close to the total value of these properties. Moreover, Mead argues that many proposed alternative bidding schemes would lead to lower total expected payments and would distort exploration and production activity by possibly favoring less attractive properties and by causing too rapid development or premature abandonment.

Chapter 8, by William Comanor, examines the efficiency of retail distribution of gasoline. Gasoline retailing has changed from the days when numerous stations offered a full range of services bundled with the gasoline purchase. More recently, the number of gasoline stations has dropped dramatically, the volume per station has increased, and incidental services are explicitly priced. Comanor reviews these trends to see if they are associated with any change in the profitability of retailing, that is, whether the consumer is getting a worse deal. He finds no evidence of changes in the degree of competition in gasoline retailing. The number of stations has, generally, declined as much for the integrated major gasoline producers as for independents. He concludes that the present structure of gasoline retailing is simply a more efficient way to deliver gasoline to the customer.

Comanor examines how the gasoline distribution system has responded to both positive and negative shocks in the price of crude oil. He finds no evidence of changes in margins that would indicate a failure of the distribution system to accommodate to changes in the wholesale cost of gasoline. His data show that service stations tend to moderate the impact of oil price shocks, for both price increases and decreases. In particular, Comanor's results do not support claims that consumers have been victims of "price gouging" after sharp changes in the wholesale cost of gasoline.

2 TRANSPORTATION ENERGY USE

David Brownstone and Charles Lave

INTRODUCTION

This chapter forecasts transportation energy demand, for both the U.S. and California, for the next 20 years. Our guiding principle has been to concentrate our efforts on the most important segments of the market. We therefore provide detailed projections for gasoline (58% of California transportation energy Btu in 1988), jet fuel (17%), distillate (diesel) fuel (13%), and residual (marine bunker) fuel (10%). We ignore the remaining 2%—natural gas, aviation gasoline, liquefied petroleum gas, lubricants, and electricity. Although we discuss prospects for the use of alternative fuels such as methanol and natural gas, we do not believe that these will be significant factors in the next 20 years. Table 2-1 gives an overview of transportation energy use in California and the U.S.

Our forecasting methodology is based on the principle that predictions should not depend on variables that are themselves difficult to predict; for example, a forecast that uses relative fuel prices as a key component is of little use if it is not possible to determine accurately the relative fuel prices. The resulting models are therefore quite simple: they depend only on such factors as demo-

The authors wish to thank Xeuhao Chu and Joe Greco for their research assistance. Richard Gilbert, Severin Borenstein, and members of UCI Transportation Lunch group provided many useful comments on an earlier draft. This research was supported in part by the University of California Transportation Center under U.S. Department of Transportation grant DTO-G-009.

[1] For expository purposes we do use forecast values of U.S. GNP in our jet fuel model, but the resulting forecasts are very similar to those from a simple time series model with a time trend.

TABLE 2–1   Transportation Energy Summary.

CALIFORNIA TRANSPORTATION ENERGY SUMMARY

California's transportation energy:
Petroleum is the source of over 99% of California's transportation energy. Transportation consumes 74% of the petroleum and 48% of the total energy used in the state.

California produces about 13% of the nation's total domestic oil.

We import half of the oil we use:
43% from Alaska and 4% from foreign sources.

75% of the oil is used in the transportation sector.

U.S. TRANSPORTATION ENERGY SUMMARY

Oil makes up 41.9% of U.S. energy.
Transportation uses 63.2% of that oil. Transportation, itself, gets 97.1% of its energy from oil. The transportation sector uses 27.3% of U.S. energy.

Of Total U.S. Vehicle Miles Traveled.
84% in autos and personal light trucks, 16% in commercial trucks.

Of Total U.S. Passenger Miles Traveled.
71% in autos, 14% in personal light trucks, 15% in commercial vehicles.

Total Freight Ton-Miles: 3,114 billion.
22% by truck, 28% by water, 19% by pipeline, 31% by rail.

Of All U.S. Transportation Energy.
40% used by cars, 18% used by light trucks, 14% used by other trucks, 3% used by off-highway vehicles, 9% used by airlines, 0.7% used by transit buses, 0.3% used by rail transit, 6% used by water freight, 4% used by pipelines, 2% used by rail freight.

WORLD TRANSPORTATION ENERGY COMPARISONS

Transportation uses more oil in the U.S.
In 1987 transportation accounted for 45% of total oil use in Europe and 38% in Japan. Public transit accounts for 22% of passenger miles in Japan, approximately 8% in Europe, and 1% in the U.S.

The U.S. had one third of all cars and buses in the world in 1988,
but this percentage is declining because cars per person is growing much faster in the rest of the world.

Gasoline prices are much lower in the U.S.
In 1989, one gallon of unleaded regular gas cost $3.41 in Japan, approximately $3.00 in Europe, and $.92 in the U.S. Nevertheless, new car fuel economy in the U.S. is similar to the rest of the world. Annual miles travelled per vehicle is about 85% of U.S. levels in Europe and 65% of U.S. levels in Japan.

SOURCES: *Economic Report of the Governor* (1990), pp. 43–45; Davis and Hu (1991), pp. xxiv, xxxi, 1–20; California Energy Commission (1992).

graphics, time trends, and airplane scrappage patterns.[1] Although our projections do not explicitly model some factors, (e.g., the effects of tightened vehicle emission standards, aircraft noise restrictions, fuel prices, and congestion), we do take them into account to the extent that these factors were present, and changing, in data from our model-calibration periods.

Our predictions are that jet and diesel fuel demand will grow at slightly lower than current rates. Gasoline demand will grow at a much slower rate because vehicle ownership is becoming saturated. We are unable to forecast residual fuel demand, but it is irrelevant for energy policy since there will be a surplus of residual fuel in California for the foreseeable future. Overall, we predict that transportation petroleum demand will grow considerably more slowly than during the last 20 years in both California and the U.S. This suggests that rapid conversion to alternative fuels cannot be justified by demand pressures.

GASOLINE

Introduction

This section projects gasoline consumption through the year 2010. We begin by projecting vehicle miles traveled (VMT), then convert this to fuel consumption using estimated average fleet miles per gallon (MPG). The VMT projection is based entirely on demographic variables: size of population age cohorts, over time; the age-based pattern of drivers' licenses, over time; and the age-based pattern of yearly VMT. At each stage, the variables are split by sex. Thus the projection method depends upon age-based and sex-based driving patterns. We will discuss the data in more detail below, but the conclusion is that we expect them to be relatively reliable.

Once we have VMT projections, we convert VMT to fuel consumption via forecasts of MPG provided by two different sources: one assumes that CAFE

---

[1] For expository purposes we do use forecast values of U.S. GNP in our jet fuel model, but the resulting forecasts are very similar to those from a simple time series model with a time trend. Except where otherwise noted, California data come from the *State Energy Data Report*, 1960-1988, published by the U.S. Department of Energy Energy, Information Administration. U.S. data come from Davis (1991), various editions of the *National Personal Transportation Survey*, and from *Highway Statistics*, published by the Federal Highway Administration.

(the congressionally mandated Corporate Average Fuel Economy standards) will remain unchanged; the other assumes that CAFE will rise from its current value of 27.5 MPG to 40 MPG by the year 2000. Our projections show the following results:

(1) U.S. population grows at 0.61% per year through the year 2010. California population grows at 1.18% per year. (2) U.S. VMT will grow at 1.94% per year through the year 2010. California VMT will grow at 2.62% per year. (3) If CAFE remains unchanged, fuel consumption will grow at 1.66% per year for the U.S., and 2.31% for California. (4) If CAFE standards are raised, fuel consumption will grow at 1.14% per year for the U.S., and 1.81% for California.

Fuel consumption grows faster in California than in the U.S., but the culprit is faster population growth, not faster travel growth; California is still receiving significant immigration. The large difference in fuel-economy standards produces relatively little difference in fuel consumption; the reason is that CAFE only affects new cars and it takes a long time for the existing fleet to turn over.

Basic Demographic Considerations

We begin by focusing on the remarkable changes in automobile availability that have occurred since World War II. In 1946 one might have spoken of the "family car" because there was approximately one car per household, and the family's many potential drivers competed for its use. But given the increase in personal income since then, and the high utility for personal mobility, families bought more and more vehicles until today we have approximately one vehicle for every potential driver. The rapid growth in the vehicle/population ratio meant that VMT, fuel consumption, and congestion all grew faster than the population.

Figure 2-1 shows the overall story: disproportionate growth of the vehicle population. The upper curve shows the size of the driving-age population. The lower curve shows the size of the personal-use vehicle fleet. Vehicles have been increasing 2.9 times faster than the population of potential drivers since 1960, and the number of licensed drivers has increased even faster.

Two demographic factors caused drivers' licenses to increase much faster than the population. First, a major fraction of the population, the baby-boomers, reached driving age during the 1960s and 1970s. Second, the enormous growth in women workers produced a disproportionate growth in women drivers. In 1947, women were only 27% of the total labor force; by 1988, they were 45% of the total labor force. But the age-transition of the baby-boomers has finished, and the growth in women's labor force participation has about reached its peak. Looking at the ratio of women workers to the total labor force, almost all the growth in this ratio occurred in the early period: it grew by 20% during the decade of the 1970s, but only 5% during the 1980s. And the U.S. Bureau of Labor Statistics predicts it will grow by only 2 percentage points during the 1990s (Fullerton, 1989). That is, the effects of these two demographic factors on the growth in demand for auto travel is about completed.

We have come through an era that produced remarkable increases in vehicle ownership and use. There will be no such changes in the future—we have nearly run these ratios to their limits. Vehicle ownership is close to saturation and the era of disproportionate growth is over.

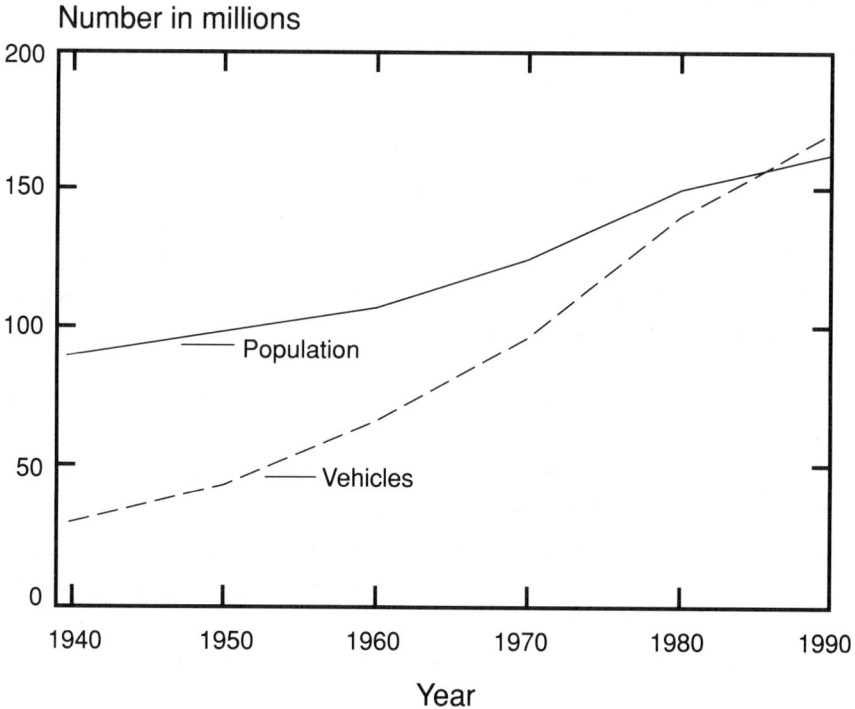

FIGURE 2-1   U.S. Driving Age Population and Personal Use Vehicles.

Projecting the Number of Licensed Drivers

The key series required to project VMT is the number of licensed drivers. We project licenses for each age/sex cohort: age cohorts are in 5-year intervals, and men and women are projected separately; thus a typical cohort might be women age 40–44. The projection requires two things: first, the number of people in each age/sex cohort, up to the year 2010, and second, the proportion of each age/sex cohort that is licensed. Population figures come from the U.S. Bureau of the Census. We make our own forecast of the proportion of licensed drivers in each cohort.

The proportion of the driving age population that is licensed has been growing steadily for as long as we have had automobiles, and is now near the point where almost everyone is licensed. Of the entire U.S. population age 15 and older: 91% of men are licensed and 79% of women are licensed, which means 85% of all the population above age 15 is licensed. Thus, a projection of the future proportion of licensed drivers has little scope for uncertainty: we are already at 85% and the theoretical ceiling is below 100%, since we must exclude 15-year-olds and the very old.

Figure 2-2 shows the licensing pattern for the U.S. in 1969 and 1989. It shows near-saturation of male licensed drivers, as expected. Two curves are shown for women: the age distribution of licensing in 1969 and 1989. Comparing 1969 to 1989, we can see that the proportion of licensed women has grown remarkably—the female age distribution curve seems to be converging on a distribution similar to the male curve.

Table 2-2 give the details for projecting the licensing pattern to the year 2010. Part A of the table shows the existing age/sex licensing pattern.[2] Part B uses simple cohort aging to project licensing patterns for future years: we assume that once licensed, a person will remain licensed. Part C then fills in the missing triangle by assuming that the licensed proportion of each successive new generation will increase by 2% per 5-year period. Finally, to allow for the effect of extreme age on the licensing rate, we project changes in the 70+ cohort based on half the percentage difference between the 70+ cohort and the

---

[2] The 56.5% figure for young males applies to the entire 15-to-19-year-old cohort (the cohort used in the Census population data). The proportion of 16-to-19-year-olds who are licensed will be about 71%.

65–69 cohort, 5 years earlier). We used a saturation limit of 100% for males and 95% for females.

Projecting VMT and Fuel-consumption

Table 2-3 shows the process for projecting U.S. total VMT and fuel consumption over the next 20 years. Part A shows projections of population size by age/sex cohorts. These projections come from U.S. Bureau of the Census Series P-17. The projection task is particularly simple in this case because they are projecting the population 15 and older, over the next 20 years: essentially all these people have already been born. They do have to project and add in immigration, but the effect on the total U.S. population will not be large because at this stage in our history growth comes mostly from births rather than immigration.

Part B repeats the drivers' license projections developed in Table 2-2. Part C shows the amount of driving by a *typical person* in each age/sex cohort. The

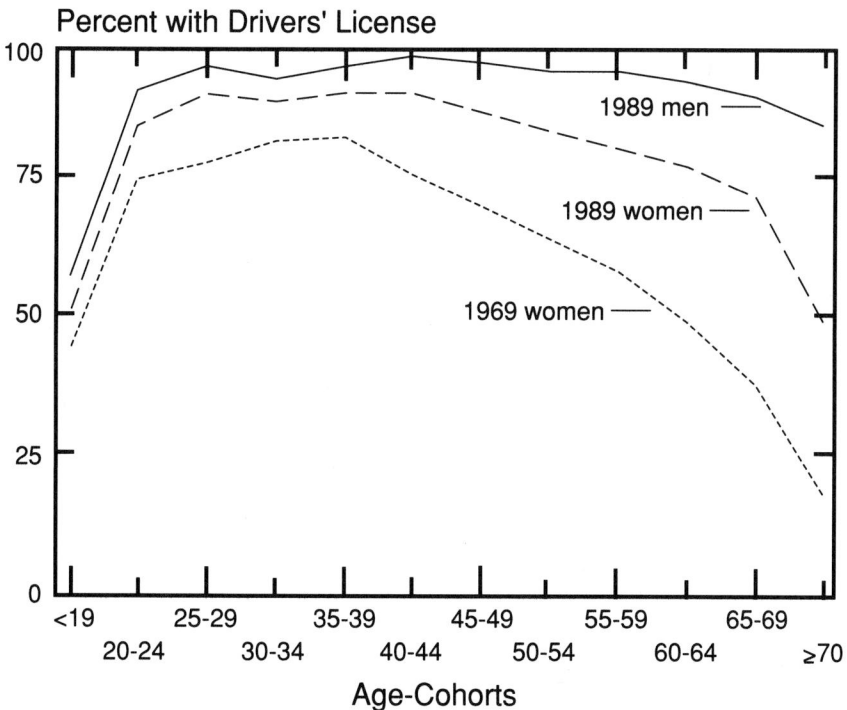

FIGURE 2-2   U.S. Drivers' Licenses as a Percentage of Driving Age Population.

TABLE 2-2   Projection of Future Age-Pattern of Drivers' Licenses.
(percentage of licensed drivers by age cohorts)

| | Males | | | | | | Females | | | | |
| | Known | — Projected — | | | | | Known | — Projected — | | | |

Part A. Projection process begins with the known distribution.

Age	1990	1995	2000	2005	2010	Age	1990	1995	2000	2005	2010
15–19	56.5					15–19	51.9				
20–24	92.7					20–24	86.4				
25–29	96.7					25–29	91.8				
30–34	94.4					30–34	90.3				
35–39	96.7					35–39	92.0				
40–44	99.0					40–44	91.9				
45–49	97.7					45–49	88.6				
50–54	96.0					50–54	84.8				
55–59	96.0					55–59	81.9				
60–64	94.2					60–64	78.3				
65–69	91.5					65–69	72.8				
70–85+	86.2					70–85+	49.9				

Part B. These tables show the effect of simple cohort-aging.

Age	1990	1995	2000	2005	2010	Age	1990	1995	2000	2005	2010
15–19	56.5					15–19	51.9				
20–24	92.7	57				20–24	86.4	52			
25–29	96.7	93	57			25–29	91.8	86	52		
30–34	94.4	97	93	57		30–34	90.3	92	86	52	
35–39	96.7	94	97	93	57	35–39	92.0	90	92	86	52
40–44	99.0	97	94	97	93	40–44	91.9	92	90	92	86
45–49	97.7	99	97	94	97	45–49	88.6	92	92	90	92
50–54	96.0	98	99	97	94	50–54	84.8	89	92	92	90
55–59	96.0	96	98	99	97	55–59	81.9	85	89	92	92
60–64	94.2	96	96	98	99	60–64	78.3	82	85	89	92
65–69	91.5	94	96	96	98	65–69	72.8	78	82	85	89
70–85+	86.2	92	94	96	96	70–85+	49.9	73	78	82	85

Part C. Next, early age cohorts grow at 2% per five-year period, 70–85+ age cohort at half the difference between prior 70–85+ cohort and prior 65–69 cohort.

Age	1990	1995	2000	2005	2010	Age	1990	1995	2000	2005	2010
15–19	56.5	58	59	60	61	15–19	51.9	53	54	55	56
20–24	92.7	95	96	98	100	20–24	86.4	88	90	92	94
25–29	96.7	99	100	100	100	25–29	91.8	94	95	95	95
30–34	94.4	97	99	100	100	30–34	90.3	92	94	95	95
35–39	96.7	94	97	99	100	35–39	92.0	90	92	94	95
40–44	99.0	97	94	97	99	40–44	91.9	92	90	92	94
45–49	97.7	99	97	94	97	45–49	88.6	92	92	90	92
50–54	96.0	98	99	97	94	50–54	84.8	89	92	92	90
55–59	96.0	96	98	99	97	55–59	81.9	85	89	92	92
60–64	94.2	96	96	98	99	60–64	78.3	82	85	89	92
65–69	91.5	94	96	96	98	65–69	72.8	78	82	85	89
70–85+	86.2	89	90	91	91	70–85+	49.9	61	64	66	67

TABLE 2-3   Process for Projecting VMT & Fuel Consumption.

	Males						Females				

Part A. Population projections from U.S. Census Series P-17, by age cohort (in thousands)

Age	1990	1995	2000	2005	2010	Age	1990	1995	2000	2005	2010
15–19	8865	8944	9735	9928	9605	15–19	8516	8585	9340	9512	9198
20–24	9244	8647	8706	9470	9648	20–24	9238	8629	8688	9432	9599
25–29	10708	9416	8808	8847	9595	25–29	10678	9424	8804	8850	9590
30–34	11195	10987	9680	9070	9108	30–34	11147	10937	9661	9034	9082
35–39	10026	11092	10882	9599	8991	35–39	10146	11105	10890	9627	9002
40–44	8691	9944	10995	10792	9527	40–44	8964	10125	11074	10863	9612
45–49	6809	8580	9822	10871	10677	45–49	7132	8903	10057	11005	10799
50–54	5590	6705	8467	9706	10748	50–54	5948	7102	8870	10029	10976
55–59	5070	5386	6478	8195	9403	55–59	5552	5842	6981	8722	9856
60–64	5032	4763	5078	6126	7770	60–64	5708	5333	5620	6720	8401
65–69	4655	4603	4382	4705	5695	65–69	5596	5453	5109	5402	6467
70–85+	8197	9199	9892	10202	10677	70–85+	13110	14508	15499	15966	16523

Part B. Percentage of each age cohort that have drivers' licenses

Age	1990	1995	2000	2005	2010	Age	1990	1995	2000	2005	2010
15–19	56.5	58	59	60	61	15–19	51.9	53	54	55	56
20–24	92.7	95	96	98	100	20–24	86.4	88	90	92	94
25–29	96.7	99	100	100	100	25–29	91.8	94	95	95	95
30–34	94.4	97	99	100	100	30–34	90.3	92	94	95	95
35–39	96.7	94	97	99	100	35–39	92.0	90	92	94	95
40–44	99.0	97	94	97	99	40–44	91.9	92	90	92	94
45–49	97.7	99	97	94	97	45–49	88.6	92	92	90	92
50–54	96.0	98	99	97	94	50–54	84.8	89	92	92	90
55–59	96.0	96	98	99	97	55–59	81.9	85	89	92	92
60–64	94.2	96	96	98	99	60–64	78.3	82	85	89	92
65–69	91.5	94	96	96	98	65–69	72.8	78	82	85	89
70–85+	86.2	89	90	91	91	70–85+	49.9	61	64	66	67

Part C. Expected number of miles per year driven by a driver of a given age

Age	1990	1995	2000	2005	2010	Age	1990	1995	2000	2005	2010
15–19	9543	10030	10541	11079	11644	15–19	7387	7764	8160	8576	9013
20–24	16784	17640	18540	19485	20479	20–24	11807	12409	13042	13707	14406
25–29	18517	19461	20454	21497	22593	25–29	11191	11762	12362	12992	13655
30–34	19592	20591	21641	22745	23905	30–34	10785	11335	11913	12521	13159
35–39	19298	20282	21317	22404	23546	35–39	11437	12020	12633	13278	13955
40–44	19396	20385	21425	22518	23666	40–44	11021	11583	12174	12795	13447
45–49	18836	19797	20806	21867	22983	45–49	9956	10464	10997	11558	12148
50–54	18081	19003	19972	20991	22061	50–54	8693	9136	9602	10092	10607
55–59	17027	17895	18808	19767	20775	55–59	7681	8073	8484	8917	9372
60–64	13308	13987	14700	15450	16238	60–64	6706	7048	7407	7785	8182
65–69	10432	10964	11523	12111	12729	65–69	5885	6185	6501	6832	7181
70–85+	8298	8721	9166	9633	10125	70–85+	3976	4179	4392	4616	4851

TABLE 2-3 (continued)

| | Males | | | | | | Females | | | |

Part D. Expected total VMT by all the drivers in a given age cohort (in billions of miles). Calculated as (population) × (driver's license ratios) × (VMT/driver's license) = total VMT.

Age	1990	1995	2000	2005	2010	Age	1990	1995	2000	2005	2010
15–19	48	52	60	66	68	15–19	33	35	41	45	47
20–24	144	144	156	182	198	20–24	94	94	102	119	129
25–29	192	181	180	190	217	25–29	110	104	103	109	124
30–34	207	219	207	206	218	30–34	109	114	108	107	114
35–39	187	212	224	212	212	35–39	107	121	126	120	119
40–44	167	196	222	235	222	40–44	91	108	122	128	121
45–49	125	168	198	224	237	45–49	63	86	102	115	120
50–54	97	124	167	197	224	50–54	44	57	78	93	105
55–59	83	93	119	160	189	55–59	35	40	52	71	85
60–64	63	64	72	92	125	60–64	30	31	35	46	63
65–69	44	48	48	55	71	65–69	24	26	27	31	41
70–85+	59	71	82	90	98	70–85	26	37	44	49	54

Part E. Total VMT

	1990	1995	2000	2005	2010
Male + Female	2,180	2,425	2,676	2,943	3,202

Part F. Average MPG of entire vehicle fleet

	1990	1995	2000	2005	2010
CalTrans Estimate	16.4	16.6	16.9	17.1	17.3
Santini Estimate	16.4	17.1	17.7	18.5	19.2

Part G. Fuel consumption by total vehicle fleet (estimated total VMT divided by estimated fleet MPG = gallons consumed)

	1990	1995	2000	2005	2010
Using Cal Trans MPG	133	146	159	172	185
Using Santini MPG	133	142	151	159	167

Part H. Comparisons: Results for year 2010 versus year 1990

	2010/1990	Annual Growth Rate (percentage)
Total Population	1.13	0.61
Driving Age Population	1.18	0.82
No. of Drivers' Licenses	1.25	1.12
Total VMT	1.47	1.94
Fuel Use, CalTrans	1.39	1.66
Fuel Use, Santini	1.25	1.14

figures come from the preliminary tabulations of the 1990 Nationwide Personal Transportation Survey. We assume that the VMT figures for each cohort will grow at 1% per year as they have in recent years.

Part D shows the *total VMT* per age/sex cohort, in billions of miles per year. It is the product of the first three matrices, A × B × C: (number of people per cohort) × (proportion of the cohort that is licensed) × (VMT per driver of given age/sex characteristics). Part E gives the aggregate totals by forecast year.

To convert VMT into fuel-consumption, we need a fuel-efficiency forecast. We use two alternative forecasts, one by Daniel Santini at the U.S. Department of Energy's Argonne National Laboratory, and one by the California Department of Transportation (Lynch and Lee, 1989).

The CalTrans projections assume stable fuel prices and no change in the federally mandated CAFE (Corporate Average Fuel Efficiency) standards that govern fuel efficiency. Despite this, the CalTrans model projects a 13% increase in average MPG of the auto fleet by the year 2010 because older autos are gradually being replaced by new ones. Similarly, CalTrans projects an 8% increase in the average MPG of light trucks by 2010. Heavy truck fuel efficiency is assumed to remain constant.[3]

The Santini projections were prepared specifically to examine the impacts of the Bryan bill now in Congress: it would mandate that CAFE be raised to 40 MPG by the year 2000, from the current level of 27.5 MPG. Santini assumes that these mandated CAFE standards are implemented. However, to the extent that the resultant new cars are less desirable to consumers—less powerful, smaller, more expensive—it will influence the *number* of new cars that are bought. Thus fleet turnover will be strongly affected because some people will keep their old cars longer. Santini incorporates these turnover effects and computes the expected average fleet MPG over time. Since 76% of light trucks are used for personal travel, he includes these in his projections as well. He estimates the fuel efficiency of the combined personal truck and auto fleet as 19.7

---

[3] Fuel costs are a significant proportion of operating costs for commercial trucks (in contrast to personal vehicles), hence commercial truckers have had very strong incentives to improve fuel efficiency since the first OPEC oil crises in 1973. CalTrans assumes that there is no room for further significant increases in the fuel efficiency of heavy trucks.

MPG in 1990 and projects that this will rise to 26.0 MPG in 2010, a 32% improvement. (These are actual, on the road, MPG figures.)

Part F shows the resultant fuel efficiencies for the total vehicle fleet (personal-use vehicles and heavy trucks), using both the CalTrans and the Santini projections. The small difference between the two projections, 19.2 versus 17.3 MPG in 2010, may seem surprising. The explanation is simply that cars are long-lived goods; it takes a very long time for the efficient new cars to replace all of the existing fleet.

Part G multiplies the fuel-efficiency projections from matrix F by the VMT projections of matrix E, to produce forecasts of total fuel consumption for the United States for the period 1990 to 2010.

Finally, Part H summarizes all the results, and puts them into perspective by comparing the changes in the important basic constituents. Over the next 20 years, our United States projections show:

(1) Drivers' licenses will grow slightly faster than the population because the transition to a fully licensed population is still going on; but that transition is nearly finished: compare the 1.12% projection with the 3.03% annual growth rate from 1950 through 1980. (2) VMT will grow at 1.94% per year, compared to the 4.62% annual growth rate for the 1950 to 1980 period. (3) Gasoline consumption will grow in the 1.14 to 1.66% range, depending upon the assumptions one makes about CAFE. Compare this to the 4.7% growth rate from 1950 through 1973.

That is, we expect a very substantial drop in the growth trends that have caused so much concern to environmentalists and conservationists.

We follow exactly the same process to make projections for California, but use California's own population structure and driving patterns. Table 2-4 shows the results and compares them to the U.S. projections. It is important to notice that California's faster growth of VMT and fuel consumption stem from its faster population growth.

Possible Influence of Public Transportation

Might increased use of public transit affect these projections? The answer is "no." Furthermore, this gloomy statement can be made with a high degree of certainty. Two main factors lead to the conclusion, and they are independent of each other; either is sufficient, by itself. First, there is little difference in energy efficiency between autos and public transit. Second, there are strong reasons to believe that it is impossible to lure a significant number of drivers onto transit.

Comparative Energy Efficiency. It may come as a surprise that there is little difference in Btu per passenger-mile between transit and automobiles. The differences in energy efficiency were never very large in the first place, and federal policy over the past 20 years has greatly reduced the gap. To begin with, federal CAFE standards have almost doubled auto fuel efficiency since 1973. In addition, as an unintended consequence of federal actions to increase transit patronage, the energy efficiency of the average transit vehicle has fallen by about 50%. (In order to make buses and trains more attractive—so as to lure drivers out of cars—federal funding encouraged conversion to air conditioned, heavier, more comfortable transit vehicles.) Table 2-5 shows the result of these changes.

TABLE 2-4   Comparison: United States vs California.

	— United States —			— California —		
	1990	*2010*	*Annual Growth Rate*	*1990*	*2010*	*Annual Growth Rate*
Total Population (millions)	250	282	0.61	29.6	37.4	1.18
Driving Age Population (millions)	196	231	0.82	22.6	30.3	1.26
No. of Drivers' Licenses (millions)	167	208	1.12	19.5	28.0	1.83
Total VMT (billions)	2,180	3,200	1.94	262	439	2.62
Fuel Use, CalTrans (billion gallons)	133	185	1.66	14.6	23.1	2.31

TABLE 2-5   Energy Intensity of Passenger Modes.
            (BTU per passenger-mile; operating energy only)

Year	*Autos*	*Buses*	*Rail Transit*	*Air Lines*
1973	5,562	2,597	2,460	8,919
1987	3,598	3,415	3,585	4,814

SOURCE: Davis and Hu (1991), pp. 2–25.

Autos, transit buses, and rail transit are now nearly equivalent. Even the airlines have made enormous gains in fuel efficiency.

Caveats: The figures are for vehicles operating with average load factors. (1) Autos used for the journey to work have lower than average load factors, so auto energy-efficiency would be decreased about 50% for that portion of auto travel (the journey to work is about 30% of auto VMT). (2) The rail figure is the average between energy-efficient old rail systems such as the New York subways, and less efficient modern rail systems like BART and the Los Angeles Metro.

Conclusion: On balance there is little difference in energy efficiency between passenger modes. Hence, to save a substantial amount of energy, we must divert a very large proportion of auto users onto transit.

The Prospects for Increasing Transit Patronage. Since 1964 the federal Urban Mass Transportation Administration has spent about $60 billion trying to find some way of luring people out of cars. The money was easily available to pay for almost any conceivable experiment: subsidized fares, free fares, newer and more comfortable vehicles, more frequent vehicle schedules, free refreshments on board, nonstop express schedules, timed transfer systems, extended operating hours, computerized scheduling, radio communication, new kinds of schedules, special fares for special groups, free parking at transit stations, advertising, image improvement campaigns, etc. None of these experiments produced significant gains in transit patronage. The federal money managed to halt the long-term decline in patronage, but they could not increase it. Transit's share of total travel has declined by more than 51% from 1960 to 1980; from 12.6% of work trips down to 6.2% of work trips, and work trips are only about a third of total travel (Pisarski, 1987, p. 48).

Radical new policy measures such as substantial parking fees would increase transit usage for the tiny proportion of travel involved in commuting to large central business districts, but the effect on the overall volume of travel would be barely measurable.

Conclusion: It's very, very hard to lure people out of automobiles and onto transit. Even if it were possible (and there is *no* evidence in the literature to support this hope), we would still not save much energy because the energy efficiency of transit and autos are roughly similar.

JET FUEL

Introduction

Jet fuel is the second largest segment of California's transportation energy use, comprising 17% of transportation energy use in 1988. This figure is almost twice the U.S. national percentage (9%) due to California's large internal air market (the Los Angeles–San Francisco corridor) and California's position as a gateway for trans-Pacific flights. The traditional method for predicting commercial aviation energy demand is to multiply U.S. Federal Aviation Administration (FAA) forecasts for available seat miles (ASM) by some measure of fleet fuel efficiency. There are a number of problems with this approach:

(1) Although the FAA collects data for ASM and revenue passenger miles (RPM) for domestic flights, these data are not available for California. There is no reason to assume that California represents a constant proportion of the U.S. national figures over a 20-to-30-year period. The FAA national statistics exclude foreign flights, which, given the large number of Pacific Rim flights originating in California, makes them less useful for determining California's jet fuel consumption.

(2) The deregulation of domestic airline service in the early 1980s is a unique event whose impact cannot be captured well by models based on time series and economic variables. One important feature of deregulation is the growth of the hub-and-spoke system, which greatly increased the number of flights to and from the hub airports as well as load factors on these flights. Of course, deregulation also lowered fares for most travelers, which also increased demand. The net effect of these changes has been to increase jet fuel consumption while greatly increasing airline passenger-miles.

(3) From an energy policy perspective, we are not interested in RPM, just energy consumed. Therefore, it is better to forecast jet fuel consumption directly. Of course, for short-haul routes such as Los Angeles–San Francisco, planes compete with automobiles, and if one mode is more energy-efficient, then the cross-elasticity between the modes may be important. However, Table 2-5 shows that the energy efficiency of the two modes was almost equal. Note that airliner efficiency increased twice as fast as other modes and should equal auto efficiency in the near future.

For these reasons, we have developed a direct forecasting model for jet fuel consumption, which depends on U.S. GNP and aircraft efficiency. This model is fully discussed in a later section. Although our projections include all non-military uses of jet fuel, we will discuss only airline passenger demand. There are some cargo-only flights, but these are a tiny fraction of scheduled airline flights. Most air cargo is still carried in the baggage compartments of passenger airlines. Although it would be interesting to compare California domestic and international airline fuel demand, we were unable to find any data sources for quantifying this split. Unlike passenger automobile demand, per capita airline demand is not saturated and thus not limited by population growth.

Policy Issues

As opposed to the situation with automobiles, California has little scope for policy intervention to change jet fuel consumption. Federal law prohibits direct state regulation, and new federal laws have also limited adding new noise restrictions on airport operations. These noise restrictions force airlines to use newer, and generally more fuel-efficient, planes. In the longer run, California can attempt to block airport construction or expansion, which will eventually restrict the growth of jet fuel consumption. In the shorter run, the main restriction on airline growth is lack of capacity in the air traffic control system, which is managed by the federal government. One California airport, Orange County, also has binding constraints on the number of takeoffs as a result of noise control litigation.

Airports generate many costs and benefits, and these features will almost certainly dominate any fuel-consumption considerations. The main costs are noise and local traffic congestion. Balancing these are the obvious benefits to businesses of proximity to airports. To minimize jet fuel consumption, the best policy would be to limit small airports and build very large airports with large ground transport feeder networks. These large airports would allow the use of larger planes with higher load factors, which would in turn increase airline fuel efficiency. Of course, the direct and indirect costs of creating mega-airports will almost certainly swamp these fuel efficiency benefits.

The easiest way to increase airline fuel efficiency is to increase the number of passengers per plane, or load factor. The development of the hub-and-spoke system during the 1980s was primarily motivated by airlines' desires to increase load factors, and they succeeded in increasing them from 55% in the mid-1970s to 63% in 1989. Further expansion of hub-and-spoke systems is limited by congestion at key hub airports. Except for small hubs at San Francisco (United) and San Jose (American), geography dictates that California will be on the spokes of major national networks. Historically, San Francisco and Los Angeles have served as transfer and refueling stops for trans-Pacific air travel. While this will no doubt continue for the foreseeable future, new long-range aircraft will permit nonstop flights from the Far East to Midwestern hub cities. It is therefore not clear that California will be involved in all the projected growth in Pacific Rim air travel.

Since a substantial portion of California air travel is on the Los Angeles–San Francisco corridor, it is conceivable that an ultra high-speed rail link would substantially reduce jet fuel consumption. Since there have not been any serious proposals for such service, it is safe to say that such a system will not be operational during the 20-year forecasting period.

Forecasting Model

We base our forecasts for jet fuel consumption on a simple linear regression model relating the log of consumption to the log of fuel efficiency (measured as ASM per gallon of jet fuel averaged over the fleet; from Greene, 1990) and the log of U.S. GNP. Although the fuel efficiency variable might appear to be purely a technological variable, it also measures the speed of replacement of old jets. The decision to replace an old, fuel-inefficient jet is strongly determined by fuel prices and demand.

Our model accounts for the unique nature of the mid-1980s period, the result of airline deregulation, by excluding observations between 1984 and 1988 while calibrating our model. We are therefore assuming that the relationships between the variables in our model will be the same in the forecast period (1989–2010) as in 1970–1984. It is important to note that our calibration period includes wide variation in GNP, jet fuel consumption, jet efficiency, and fuel prices. Our model does not explicitly consider the effects of increasing fuel prices and incomes because these items are very difficult to determine. Of course, income is strongly related to GNP, which is included in our model.

Figure 2-3 shows jet fuel consumption for California and the U.S. California's higher rate of increase in the 1980s may be due to the increase in travel between the U.S. and the Pacific Rim. A large fraction of this travel involves jet fuel purchases in California. Comparison of figures 2-3 and 2-4 show that the number of passengers carried by domestic airlines has grown much faster than fuel consumption, due to the large increase in fleet fuel efficiency (Figure 2-5) over the period. Note that Figure 2-5 shows only the fuel efficiency due to the use of newer, more fuel-efficient aircraft. Another way to carry more passengers without using more fuel is to increase load factors, which is one of the main effects of airline deregulation. Our forecasting model for California is given by:

log(billion gallons of jet fuel) =
   13.6 − .8 × log of fuel efficiency + .86 × BUSCYC + .033 × Time
   (R^2 = .56)

where we have decomposed the logarithm of U.S. GNP into a time trend (Time) and a pure business cycle measure (BUSCYC), and fuel efficiency is ASM per gallon of jet fuel averaged over the U.S. jet fleet. This model is estimated over 1970 to 1984, and all of the coefficient estimates are significant at the 10% level. Holding efficiency and business cycle effects constant, the model predicts a 3% annual growth in California jet fuel demand. Holding GNP and time constant, a 1% increase in jet fleet fuel efficiency is associated with a 0.8% drop in jet fuel consumption.

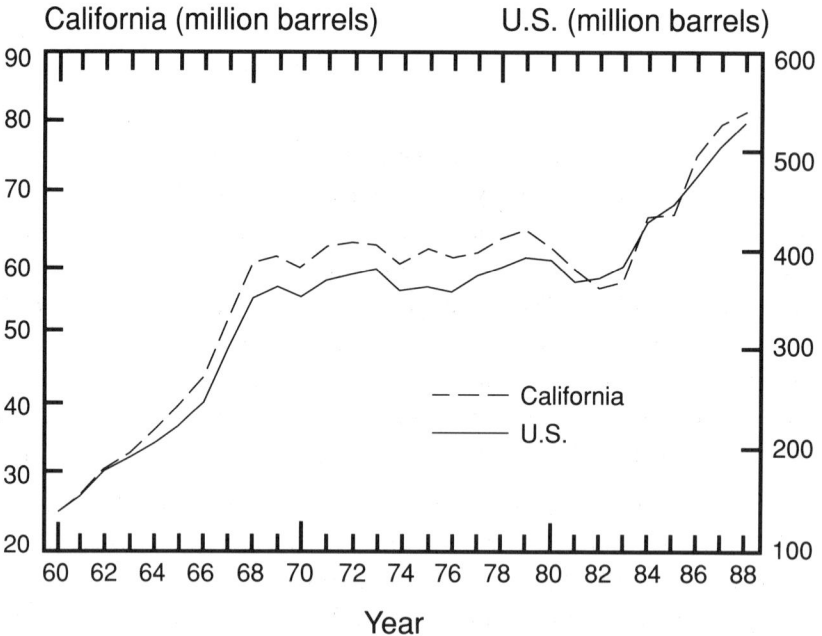

FIGURE 2-3   California and U.S. Annual Jet Fuel Consumption.

RPM in billions

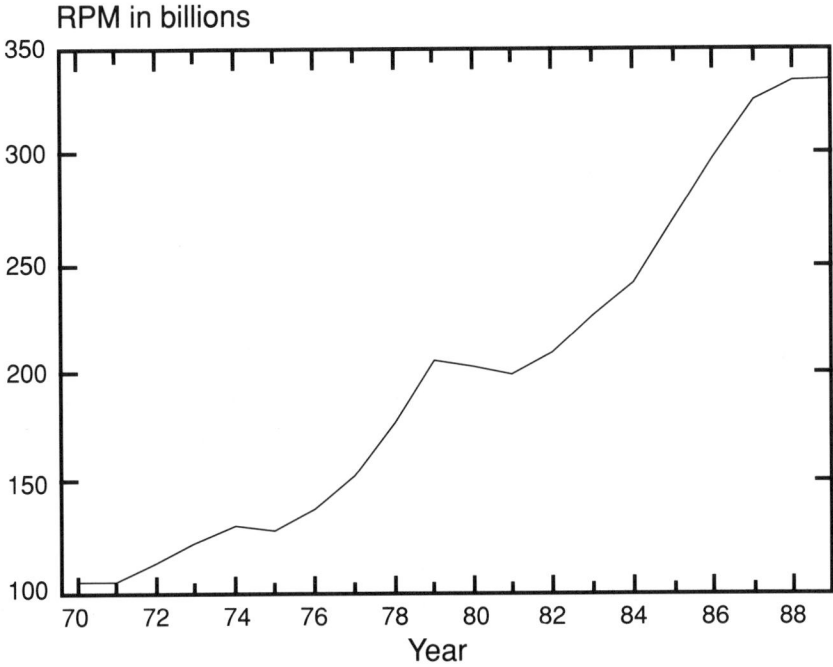

FIGURE 2-4   U.S. Airline Annual Revenue Passenger Miles.

Available Seat Miles per Gallon

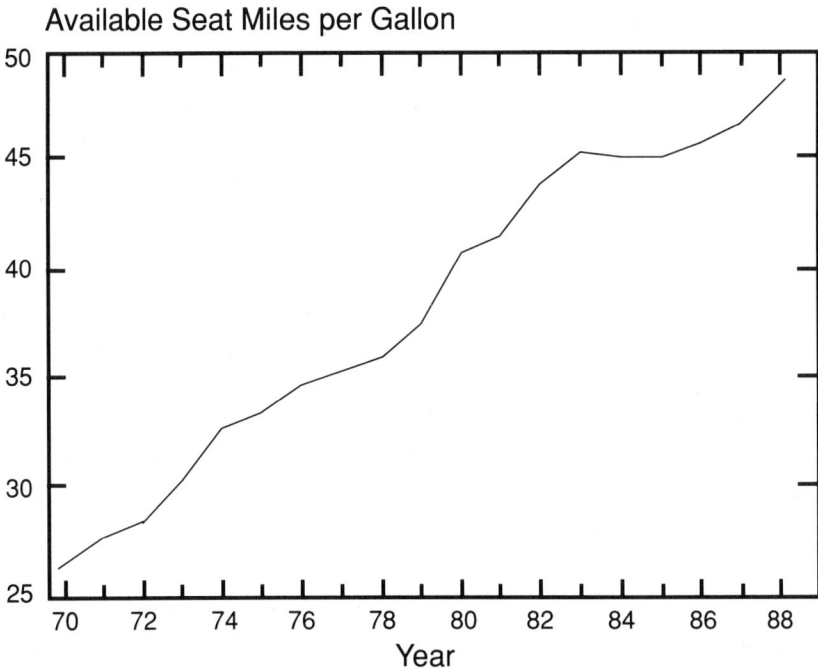

FIGURE 2-5   U.S. Jet Fleet Fuel Efficiency.

A similar forecasting model for the U.S. is given by:

log(10 billion gallons of jet fuel) =
    9.1 − log of fuel efficiency + .45 × BUSCYC + .047 × Time
    (R^2 = .70)

Relative to California, the U.S. model shows a higher time trend and slightly higher sensitivity to fuel efficiency. Assuming that U.S. GNP and fuel efficiency are known, both of these models give accurate predictions. For example, the forecast jet fuel consumption for the year 2000 will be between plus and minus 15% of the forecast value with 90% probability.

To determine future fuel consumption from this model, we need projections for the independent variables, U.S. GNP and jet fuel efficiency. For 1990–2000 we used the FAA forecasts for GNP, which are based on a consensus forecast from Wharton and DRI. For 2001–2010 we used our own projections derived from regressing the log of GNP on time. The resulting series has real GNP growing at a 2.6% rate over the period 1970–2010. For fuel efficiency, we used two scenarios from Greene (1990). The base case assumes no retrofitting (primarily new engines) and no "new generation" aircraft through 2010. Efficiency improvements still occur in the base case due to retirement of old, inefficient aircraft that are replaced by more efficient current models. The efficient scenario assumes new generation aircraft available in 2000 together with accelerated scrappage and retrofitting of old planes. Greene views these cases as extremes bracketing the likely actual values.

To account for the unique 1984–1988 deregulation period, we produce forecasts for 1985–2010, and adjust all the figures upward so that the forecast equals the actual value for the last year of real data, 1988. This adjustment results in a 4% upward adjustment of our forecast values. This method treats the 1984–1988 increase in airline passenger and fuel demand as a unique event, caused by lower deregulated airfares and the switch to hub-and-spoke domestic networks. We are therefore assuming that for the next 20 years (1989–2010) the relationship between airline fuel demand, GNP, and airplane efficiency will follow the same patterns as in 1970–1984. Since the 1970–1984 period includes wide changes in business cycles and fuel prices, our forecasts should be valid as long as future variation in these variables is not much greater than in the 1970–1984 calibration period.

The projections (Figure 2-6) for the two different fuel efficiency scenarios do not differ significantly until 2000, with both showing a 25% increase in California jet fuel consumption over the 1988 base year. For the 2001–2010 period, however, the forecasts diverge. The base case shows a continued 20% increase, while the efficient case shows only a 5% increase. Which of these two scenarios is more likely depends largely on fuel prices. The faster jet fuel prices increase significantly, the more likely it is that jet fleet fuel efficiency will follow the efficient scenario. Figure 2-7 shows forecasts for U.S. jet fuel consumption using the same methods. These forecasts are similar to California's, except that U.S. jet fuel consumption is predicted to grow at a slightly faster rate.

Conclusions

California jet fuel consumption will rise by approximately 25% during the 1989–2000 period, followed by a slower increase in 2001–2010. Key factors affecting jet fuel consumption are U.S. GNP, fuel prices, and aircraft fuel efficiency. There are no reasonable policies California can pursue to significantly alter these factors. In the short run, California has limited scope for changing other factors such as airport congestion and aircraft noise limitations because these are largely regulated by the federal government. In the very long run (more than 20 years), policies that shift passengers in the Los Angeles–San Francisco Bay Area corridor to surface modes will reduce California jet fuel consumption, but it is not clear that this will reduce total energy consumption.

DISTILLATE FUEL

Introduction

Distillate fuel for transportation use is primarily diesel fuel used for trucks, railroads, and ships. Distillate fuel accounts for 13% of the transportation Btus in California for 1988 and 20% in the U.S. In 1981, the last year for which data for California are available, 14% of distillate fuel was used by the military, 18% by railroads, 5% by ships, and 63% for on-highway use. There are no statistics for the breakdown of on-highway use for California, but for the U.S. in 1981, trucks consumed 90% of on-highway distillate fuel, cars 5% and buses the remaining 5%. By 1988 these U.S. breakdowns remained the same, except that rail's share declined to 10% and on-highway share increased to 67%.

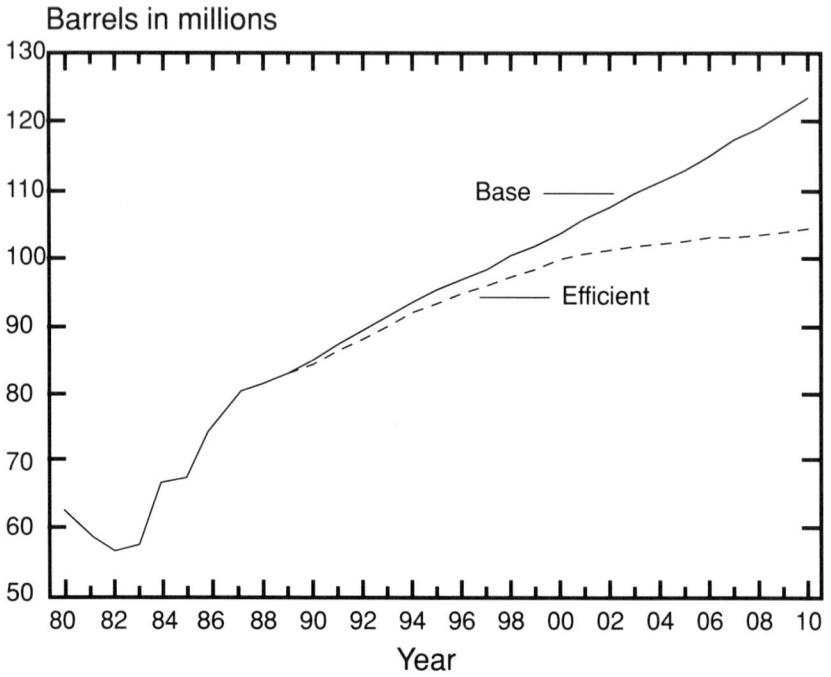

FIGURE 2-6   California Annual Jet Fuel Consumption Forecast.

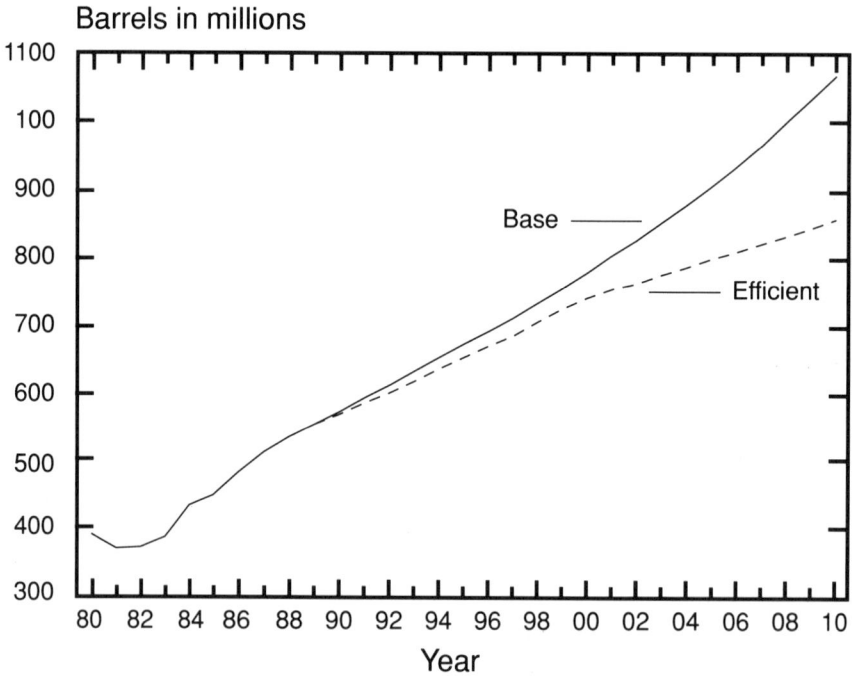

FIGURE 2-7   U.S. Annual Jet Fuel Consumption Forecast.

Given the disparate transportation uses for distillate fuels and the lack of detailed California data on the separate uses, we will use a simple time series model to generate our forecasts. Note that the large majority (97%) of distillate (diesel) fuel is used to haul freight, with trucks, rail, and ships all competing for business. While trucks and rail are almost totally fueled by diesel, all but the smallest ships are fueled by residual fuel oil, which is discussed in the next section of this report. Diesel fuel has also been a competitor with gasoline as a fuel for cars and light trucks. This tradeoff will be discussed further in the policy discussion below, but it appears that emission limits on cars and light trucks will limit diesel use for these vehicles.

Since our forecasts are based on the assumption that California distillate fuel consumption will continue to follow the same historical trends as in the 1964–1988 period, the next subsection will concentrate on evaluating policies that might invalidate this key assumption. The last subsection will present the forecasting model and its results.

Policy Issues

Although there have not been any suggestions that a new generation of fuel-efficient diesel engines is emerging, there has been considerable policy interest in reducing emissions from diesel trucks and buses. The recent federal clean air act mandates reductions in particulate and other diesel emissions beginning in 1994. One of the more popular technologies for reducing diesel emissions, especially for urban transit buses, is to convert the engines to run on methanol. Of course, if there are many such conversions then our projections for future diesel fuel consumption will be too high. Although there does appear to be some evidence that either pure or 85% methanol mixtures will yield substantial ozone reductions for gasoline engines (Walls and Krupnik, 1990), the evidence on methanol's ability to reduce diesel emissions is mixed.

There have been a number of trials with standard transit buses converted to run on methanol, and the Southern California Rapid Transit District (SCRTD) in Los Angeles has just begun testing of transit buses with diesel engines specially designed to run on methanol. The results of the earlier tests are summarized in Santini and Rajan (1990). Small (1988) and Small and Frederick (1989) perform cost-effectiveness and cost-benefit studies for methanol buses and particulate traps. Although constant-speed dynamometer tests show substantial emission reductions, later tests under more realistic stop-and-go conditions show no decrease in emissions. Methanol with a platinum catalyst can reduce particulate and hydrocarbon emissions, but it does not significantly decrease nitrogen oxides (NO_x). Worse yet, methanol buses emit much higher levels of formaldehyde. Even if emissions are reduced, there are still unknown additional costs associated with increased maintenance and reliability problems relative to standard diesel engines.

Given the reality of the 1994 emission controls, diesel manufacturers are actively pursuing other technologies for meeting the standards. According to *Metro Magazine*, July/August 1989, Volvo and Iveco have both produced prototype combination particulate traps and catalytic converters, which allow current diesel engines to meet the 1994 standards. Detroit Diesel, a major American manufacturer, is also developing particulate traps to be tested on New York City buses. Although the reliability of these systems is unknown, the fact that they can be added to existing engines suggests that the overall vehicle will be more reliable than new methanol diesels. Manufacturers claim that these particulate traps do not affect engine performance or fuel efficiency. If this is true, then adoption of these particulate traps will also not affect our forecasts. The main effect of these traps will be to raise the capital costs of buses and trucks, which in turn will tend to make their operators less sensitive to fuel price changes.

Another possibility, currently being tested in Sweden, is to add steps to the refinery process to clean up diesel fuel. Preliminary results suggest that this clean, low-sulfur fuel combined with standard catalytic converters can also meet emission standards. This technology will, of course, increase the price of diesel fuel, which would tend to make our figures too high. Cleaning up fuel, which is mandated for gasoline by 1995 in the new clean air act, is an appealing policy because it reduces emissions from all vehicles, not just new ones. Small and Frederick (1989) find that although adoption of methanol buses can lead to higher emission reductions, the per unit costs of these reductions are much higher than with cleaner fuel or particulate traps. Their analysis ignores the potentially high maintenance and reliability costs associated with new methanol diesel engines.

Economic efficiency strongly suggests that it is better to set standards and let the marketplace choose the best technology rather than dictate which technology to use. California's policies have promoted methanol as a partial solution to air quality problems. Our review of the current technologies suggest that methanol may be a costly choice for cleaning up diesel engines.

Another policy issue that has received less recent attention is the competition between surface freight modes. For the U.S. in 1988, average truck fleet efficiency was 3460 Btu per ton-mile while water used only 361 Btu per ton-mile and rail 434 Btu per ton-mile. Therefore, if one is interested only in reducing transportation energy use, it is best to shift freight from trucks to either rail or water. Unfortunately, the deregulation of the trucking industry in the late 1970s lowered the relative price of truck transport and therefore increased diesel fuel consumption (Winston, Corsi, Grimm, and Evans, 1990). If the only objective is to reduce fuel consumption, then the efficient solution to this problem is to deregulate rail and charge truckers the full costs of providing the interstate highway system services. This solution is clearly infeasible with current fuel prices, but if it occurred, diesel consumption would be reduced relative to our forecasts. Another option is to subsidize rail service, but Winston shows that the required subsidy levels are politically unrealistic. If domestic water shippers were forced to compete on the same "level playing field" basis (paying full costs for channel dredging and port facilities), then they would probably lose business to rail.

There was a large increase in diesel car and light truck sales in the years immediately following the 1979 oil price shock. The reason for the popularity of diesel in this period can be seen in Figure 2-8, which shows the price of diesel fuel and unleaded regular gasoline. When the price of unleaded gas dropped in 1984, sales of diesel cars rapidly dropped to almost zero. Even if equivalently large price differences develop in the future, it is unlikely that there will be a resurgence of diesel vehicle purchases because of their difficulty in meeting emission standards.

Air quality concerns have led to the consideration of compressed natural gas (CNG) as a fuel for cars and light trucks. Although CNG definitely reduces emissions, high distribution costs make it unlikely to be used for anything other than centrally fueled fleets in the near future.

Forecasting Model

Since most diesel fuel is used for hauling freight, and since freight movement should be closely related to economic activity, we begin with a simple model

¢/10,000 Btu

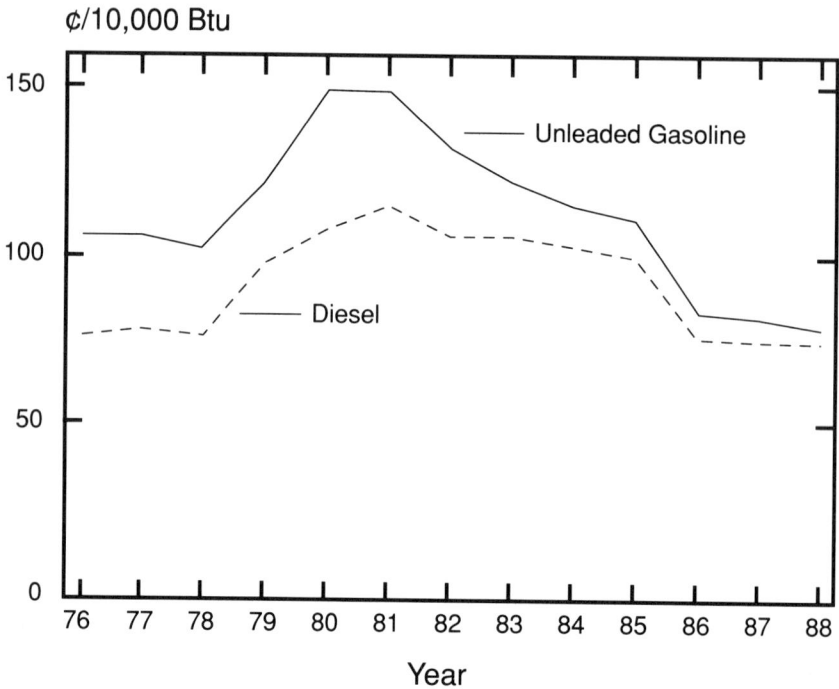

FIGURE 2-8   U.S. Retail Prices for Motor Fuels.

relating the log of annual California distillate fuel consumption in millions of barrels (LCADIF) to a 1 year lag in LCADIF and the log of California Gross State Product (GSP) in 1982 dollars (LCGASP82). The estimated model using observations from 1964 to 1986 is given by:

LCADIF =
.6 × LCADIF(1 year earlier) + 1.18 × LCGASP82 − 4.54
(R^2 = .94)

Figure 2-9 shows the time series plots of the raw series, and Figure 2-10 shows that a similar relationship holds for the entire U.S. as well. These results confirm the strong positive correlation between distillate fuel consumption and economic activity.

One difficulty with using the above model is that it requires good predictions for California GSP. We were unable to find a long enough consistent series to generate such a forecast, so we then tried replacing California GSP

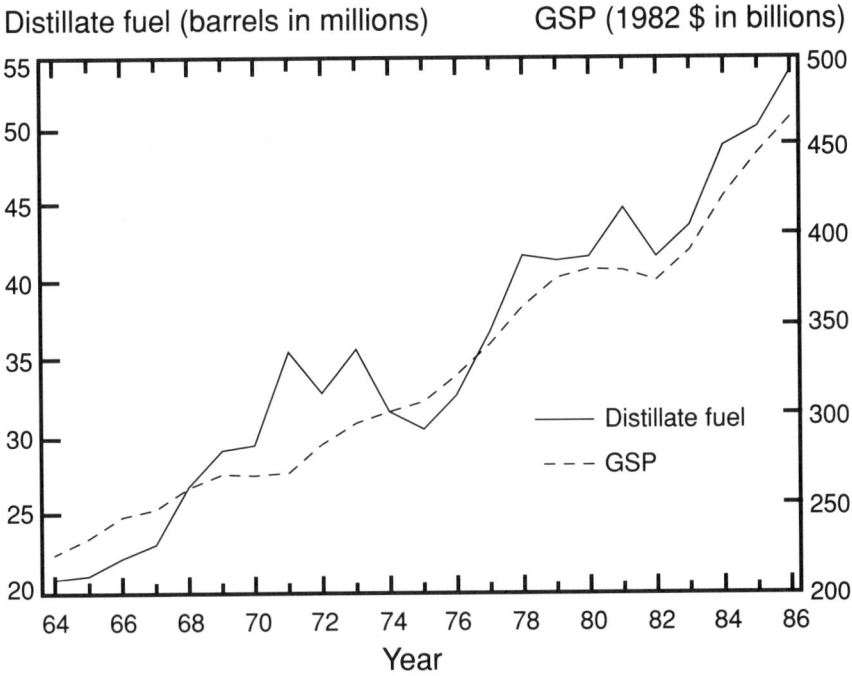

FIGURE 2-9   California Annual Distillate Fuel Consumption and Gross National Product.

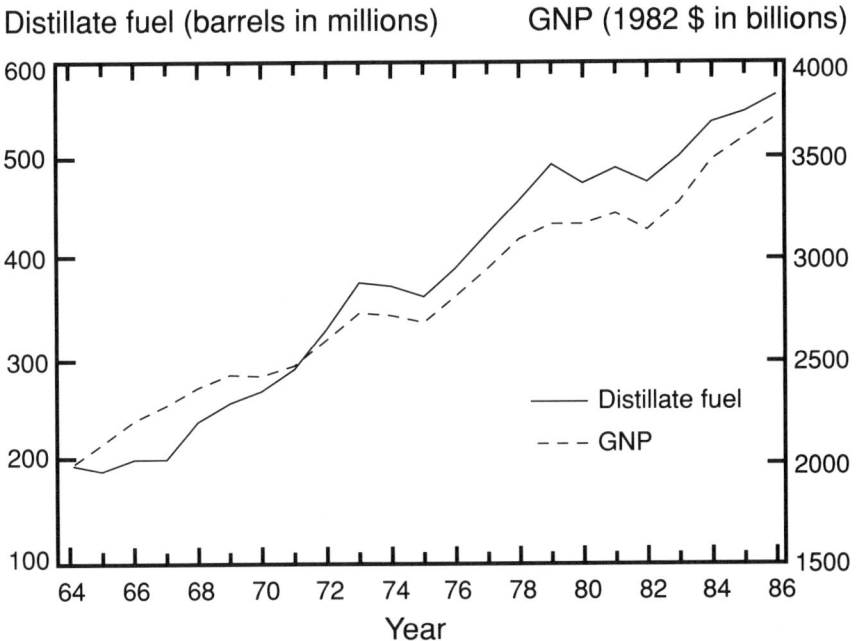

FIGURE 2-10   U.S. Annual Distillate Fuel Consumption and Gross National Product.

with U.S. GNP. This model did not fit as well as an even simpler time series model relating LCADIF to a 1-year lagged value. Our actual forecasts are generated from this model fit over the period 1960–1988, and they are shown in Figure 2-11 along with similar forecasts for the entire U.S. These figures show a 13% growth in California over the 1990–2000 period, followed by 8% growth over the 2000–2010 decade. Because the last year of data, 1988, corresponded to unusually high distillate fuel consumption, the level of our forecasts is probably high. U.S. distillate fuel use has been more stable, and our projections show slightly higher growth than in California.

Conclusions

We predict continued moderate growth in California distillate fuel consumption during the next 20 years. Most of the foreseeable reasons why our forecasts could be wrong suggest that they will be too high. Nevertheless, the scenarios leading to significant reduction in diesel fuel consumption are not likely to occur, especially during the 1990–2000 period. The largest unknown factors are future fuel prices and future technology for reducing diesel emissions.

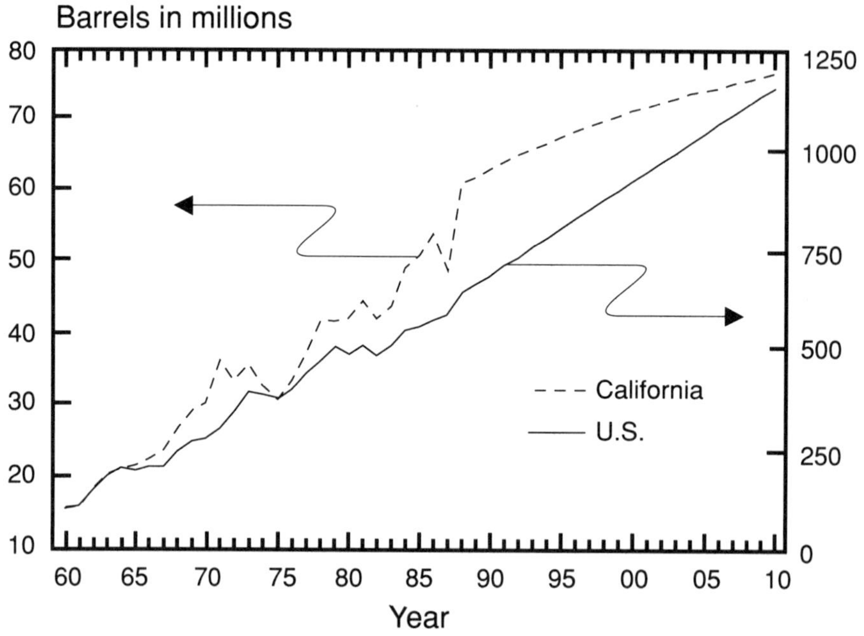

FIGURE 2-11   California and U.S. Annual Distillate Fuel Consumption Forecasts.

RESIDUAL FUEL

Residual fuel, or bunker fuel, is the very heavy oil left over after the refining process. It is used only in ships and power plants, but emission control regulations prohibit its use in California power plants. Figure 2-12 shows that residual fuel consumption fluctuates widely and follows no discernible pattern for either California or the U.S. California accounts for approximately 30% of U.S. residual fuel consumption. Although this makes forecasting difficult, from an energy policy perspective we need not worry about residual fuel. Most of the information for this section came from personal communications with Tom Burns, head of the Economics Department, and Dick Parmalee, head of Marine Fuel Marketing, at Chevron, a major player in the market.

   The total world demand for bunker fuel is stable, smooth, and predictable. A typical large ship can hold enormous amounts of fuel (displacing ballast, if necessary), which allows operators great freedom in choosing fueling locations. Since fuel costs are a large fraction of ship operating costs, ship operators

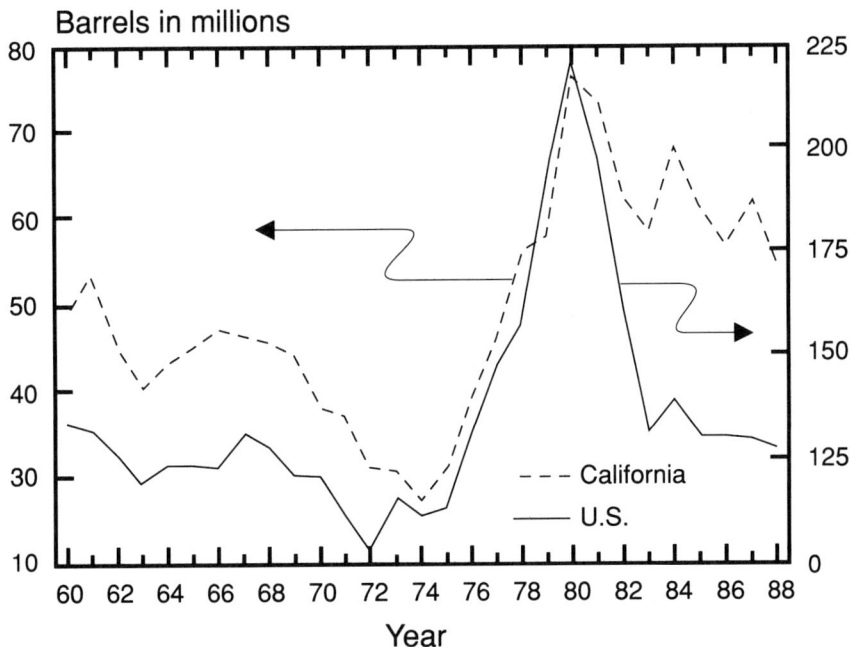

FIGURE 2-12   California and U.S. Annual Residual Fuel Consumption.

have a strong incentive to look hard for the lowest fuel price. The amount of bunker fuel sold out of California will therefore be extremely sensitive to relative prices between California and alternative ports, which accounts for the large swings in Figure 2-12. Note that since a large fraction of California shipping is from Asia or through the Panama Canal, the set of alternative ports includes almost the entire world. It is also possible for a ship to take on enough fuel in Indonesia (or California, if cheaper) for a round-trip journey across the Pacific.

From an energy policy perspective, none of this matters very much because California has a large surplus of bunker fuel, which will last for many years. California crude oil is very heavy, while California demand is for products, such as gasoline, more easily refined from light crude. With current refining technology, much of a typical barrel of California crude cannot be used in California. Depending on relative prices, this surplus is either shipped east or sold as bunker fuel at California ports. Therefore California will have more than enough bunker fuel available at world prices to service the West Coast shipping trade, which is a small fraction of shipping through our ports.

California refiners would like to reduce this surplus by modifying their refineries to get more profitable gasoline out of a barrel of California crude oil. The current technology for doing this, called "cokers," also increases air pollutant emissions from these refineries. Therefore, unless some new technology emerges soon, it is unlikely that air quality standards will allow much reduction in California's residual fuel surplus.

Since there is no energy policy reason to care about California residual fuel demand, we have not produced any numerical forecasts.

REFERENCES

Asin, R. H. (1980). "Characteristics of 1977 Licensed Drivers and Their Travel." Nationwide Personal Transportation Study, Report No. 1. Washington, D.C.: U.S. Department of Transportation, Federal Highway Administration.

California Energy Commission (1992). The Biennial Energy Report. Sacramento.

Davis, S. C., D. B. Shonka, G. J. Anderson-Batiste, P. S. Hu (1989). "Transportation Energy Data Book," Edition 10. Oak Ridge National Laboratory Report ORNL-6565. September. Springfield, Va.: National Technical Information Service.

Davis, S. C. and P. S. Hu (1991). "Transportation Energy Data Book," Edition 11. Oak Ridge National Laboratory Report ORNL-6649. January. Springfield, Va.: National Technical Information Service.

Fullerton, H. N., Jr. (1989). "New Labor Force Projections, Spanning 1988-2000." *Monthly Labor Review*, Vol. 112, No. 11, pp. 3-12.

Greene, D. (1990). "Energy Efficiency Improvement Potential of Commercial Aircraft to 2010." Oak Ridge, Tenn.: Center for Transportation Analysis, Oak Ridge National Laboratory. June.

Lynch, R. A. and L. F. Lee (1989). "California Motor Vehicle Stock, Travel and Fuel Forecast." November. Sacramento: California Department of Transportation.

Pisarski, A. E. (1987). *Commuting in America: A National Report on Commuting Patterns and Trends*. Westport, Conn.: The Eno Foundation for Transportation, Inc.

Santini, D. J. and J. B. Rajan (1990). "Comparisons of Emissions of Transit Busses Using Methanol and Diesel Fuel." *Transportation Research Record*, Vol. 1255, pp. 108-118.

Small, K. A. (1988). "Reducing Transit Bus Emissions: Comparative Costs and Benefits of Methanol, Particulate Traps, and Fuel Modification." *Transportation Research Record*, Vol. 1164, pp. 15-22 (1988).

Small, K. A. and S. J. Frederick (1989). "Cost-effectiveness of Emissions Control Strategies for Transit Buses: the Role of Photochemical Pollutants." *Transportation Research*, Vol. 23A, pp. 217-227.

Walls, M. A. and A. J. Krupnik (1990). *Resources for the Future Newsletter*, Summer.

Winston, C., T. M. Corsi, C. M. Grimm, and C. A. Evans (1990). *The Economic Effects of Surface Freight Deregulation*. Washington, D.C.: Brookings Institution.

天然66

3 CRUDE OIL SUPPLY AND DEMAND

Walter Mead

INTRODUCTION

The industrial world economies run on oil. When will the supply of crude oil run out? How will markets react as evidence ultimately appears in the form of rising oil prices that world crude oil reserves and resources are in a persistent decline phase? How will ultimately declining oil supplies affect the United States, and sub-regions including the West Coast and California? These are the questions to be addressed in this chapter.

To develop answers to these questions we will survey what is known and estimated about world, U.S., and West Coast (California) crude oil reserves and resources. From an economic perspective, the early evidence of resource depletion will appear in the form of higher costs of finding and producing crude oil, followed by the introduction and expansion of substitutes for conventional crude oil. Consequently, we will review evidence of crude oil production costs and supply costs for the least costly oil substitutes.

The crude oil market is an international market with relatively few tariff or quota barriers to trade. Therefore, the United States, the West Coast, and California are sub-markets of a single international market, with crude oil prices differing only by transportation costs and minor barriers due to government interference.

Walter Mead is with the Department of Economics, University of California, Santa Barbara.

WORLD CRUDE OIL RESERVES, PRODUCTION, AND CONSUMPTION

We will examine data on world oil reserves and likely future discoveries in order to appraise likely future oil production over the next two decades. Both exploration and production are heavily influenced by actual and expected crude oil prices. Price, in turn, is determined by the interaction of supply and demand, given assumptions about economic growth, population growth, changes in exploration and recovery technology, development and cost of oil substitutes, and other economic forces.

A word of caution about economic forecasting

Oil supply, demand, and price are subject to unforeseen and unforeseeable developments that produce major changes, especially in price. Both demand and supply are relatively inelastic in the short run. Consequently, any sudden change in either supply or demand can produce major changes in price, which, in turn, affect future supply and demand.

In 1980 and 1981, the Energy Modeling Forum (EMF) at Stanford University brought together what the EMF management felt were the most competent energy forecasting modelers to provide forecasts of the market clearing price of crude oil through the year 2020. With 10 years of hindsight, it is now possible to evaluate the performance of these forecasts during their first decade.

The actual price of crude oil in real terms (1988 dollars) declined sharply from 1980 through 1991. In contrast, the forecasts indicated either stable or sharply rising prices. Figure 3-1a,b shows the forecasting errors. The nine forecasters failed to anticipate two factors. First, an expansion in exploration and production provided by the incentive of higher real prices in the decade of the 1970s, in particular, the non-OPEC supply response, was underestimated. (For a detailed analysis of this point, see Porter, 1990.) Second, of less importance, demand response to higher prices in the 1970s appears to have been underestimated. This record produces humbling results. While forecasting is an inexact science, oil price forecasting is particularly hazardous.

World crude oil reserves

Crude oil reserves are dominated by a few "super-giant" fields, defined as fields larger than 5 billion barrels. There are only 38 super-giant fields in the world, but they originally held more than 50% of the known world reserves.

Index (1980 = 100)

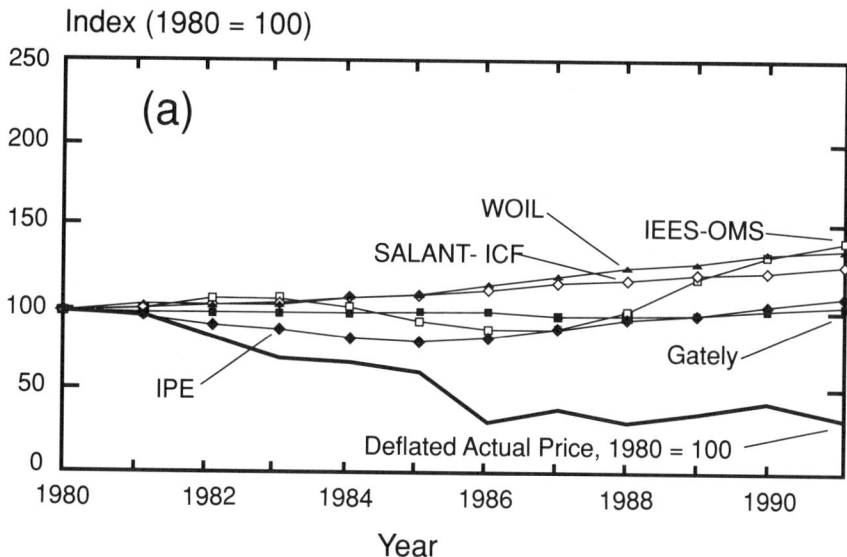

(a)

WOIL
IEES-OMS
SALANT- ICF

IPE

Gately

Deflated Actual Price, 1980 = 100

Year

Index (1980 = 100)

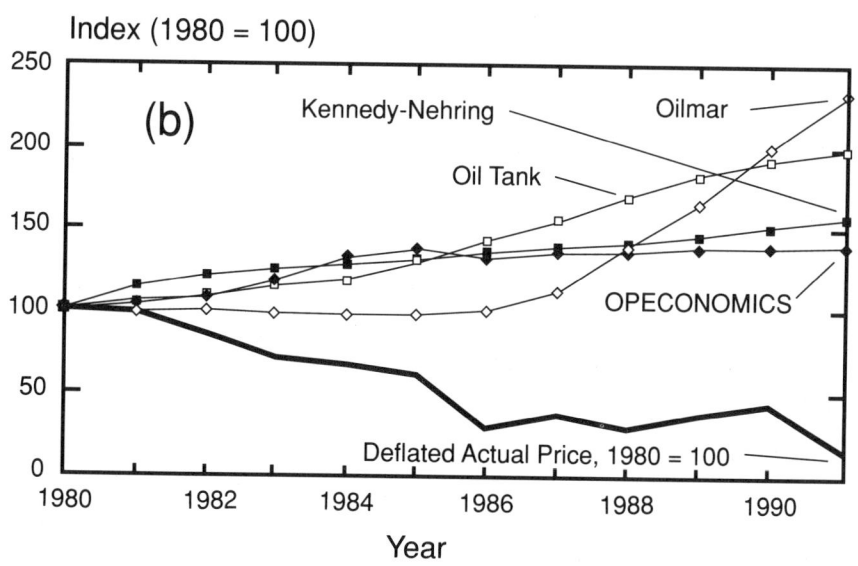

(b)

Kennedy-Nehring
Oilmar

Oil Tank

OPECONOMICS

Deflated Actual Price, 1980 = 100

Year

FIGURE 3-1   Crude Oil Price Forecasts vs Actual Price. SOURCES: a) Gately =
Gately-Kyle-Fischer, New York University and Imperial Oil, Ltd.; IEES = Calvin Kilgore,
International Energy Evaluation System–Oil Market Simulation, U.S. Department of
Energy; IPE = Nazli Choucri, International Petroleum Exchange, MIT; Salant-ICF =
Stephen Salant and William Stitt, U.S. Federal Trade Commission and ICF, Inc.; WOIL =
John Stanley-Miller, U.S. Department of Energy and Energy and Environmental Analysis,
Inc. b) Kennedy-Nehring = Michael Kennedy and Richard Nehring, University of Texas
and Rand Corp.; Oil Tank = Leif Ervik, Michaelsen Institute; OPECONOMICS = John
Mitchell, British Petroleum Co., Ltd.; OILMAR = Frank Potter, Energy and Power
Subcommittee, U.S. House of Representatives.

Twenty-six of these fields are in the Persian Gulf region. Another 30% of world reserves are accounted for by about 300 giant fields (500 million to 5 billion barrels). About 1,000 fields ranging in size from 50 million to 500 million barrels accounted for another 15% of the reserve.

There are about 40,000 oil fields in the world. Therefore, less than 5% of the world's oil fields originally contained over 95% of the total world reserves (Riva, 1991, p. 1). "Since the geology of virtually all of the world's sedimentary basins is at least partially known, geological inference indicates that it is unlikely that any undeveloped region will be found to contain more recoverable oil than is currently produced in one year, 22 billion barrels" (*Ibid.*, p. 2).

Currently, 66% of the world's crude oil reserves are in the Middle East, but only 26% of the world's crude production is from that area. Non-Middle Eastern crudes are being produced at nonsustainable rates in order to gain a measure of energy independence, or to maximize current gross income, as in the case of the Commonwealth of Independent States (CIS).[1] A degree of independence today is sought at the expense of severe dependence later.

Mineral "reserves" are defined as known resources that are producible at current prices and with current technologies. The word "resource" is used to describe minerals that are currently submarginal. When crude oil prices increase, or technological improvements lower production costs, some sub-marginal resources become economically viable and are shifted from the resource to the reserve category. Furthermore, higher product prices and technological advances lead to more exploration, which shifts resources from the unknown to the known category, some of which will be economically viable and become reserves.

The record of crude oil reserves is shown in Table 3-1 and Figure 3-2. These reserves have grown more than ten-fold from 1948 through January 1, 1991. The sharp increases in 1988 and 1990 are due almost entirely to restatements of reserves by Iran, Iraq, Saudi Arabia, and the United Arab Emirates.

Under the unreal assumptions that no new oil reserves will be found and consumption levels remain constant, world reserves are sufficient for 45 years. However, it is obvious that new reserves worldwide have been found faster than they have been produced for the years shown in Table 3-1. While this record cannot continue into the indefinite future, it is to be expected for several additional decades. The four Middle Eastern producers noted above have re-

---

[1] Formerly the Soviet Union.

TABLE 3-1   Estimated Proven Reserves of Crude Oil. (million barrels)

Year	California, West Coast[a]	United States	Total World
1948		21,488	68,198
1949		23,280	73,599
1950		24,649	76,453
1951		25,268	89,926
1952		27,468	103,444
1953		27,960	116,454
1954		28,945	134,970
1955		29,561	157,500
1956		30,012	189,571
1957		30,435	227,958
1958		30,300	260,640
1959		30,536	272,403
1960		31,719	290,035
1961		31,613	297,744
1962		31,758	305,407
1963		31,389	309,514
1964		30,970	326,947
1965		30,991	338,668
1966		31,352	348,119
1967		31,452	381,885
1968		31,377	407,553
1969		30,707	454,735
1970		29,632	530,534
1971		39,001	611,195
1972		38,063	632,553
1973		36,339	664,220
1974		35,300	626,706
1975		34,250	712,419
1976		32,682	657,921
1977	5,005	33,502	642,639
1978	4,974	31,780	647,999
1979	5,265	31,356	645,323
1980	5,470	27,051	642,175
1981	5,441	29,805	651,930
1982	5,405	29,426	670,350
1983	5,348	27,858	668,262
1984	5,707	27,735	669,738
1985	5,801	28,446	699,813
1986	5,708	28,416	700,567
1987	5,746	26,889	697,450
1988	5,903	25,270	887,348
1989	5,803	26,500	907,443
1990		25,860	1,001,572
1991		26,177	999,113

SOURCES: Oil and Gas Journal (1990b); U.S. Deaprtment of Energy (1990b).
[a] As of January 1 of the year.

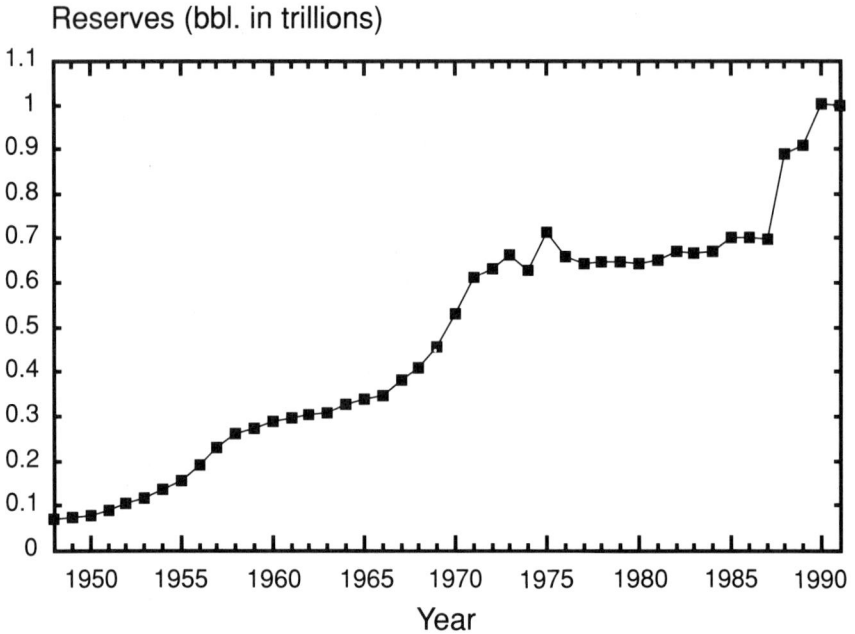

FIGURE 3-2   Total World Crude Oil Reserves.

serves so large that there has been no economic incentive for them to search for new oil fields. For example Saudi Arabia, with 258 billion barrels of oil reserves, can produce for 85 years even at the peak Persian Gulf War levels of production (8 million barrels per day). At pre-war levels of production the Saudi reserves are sufficient for about 170 years. With very high reserve-to-production levels, an investment in exploration in Saudi Arabia would have a very low rate of return.

Relative to the assumption that consumption levels remain constant, it is apparent that oil consumption growth rates depend on real oil prices, along with economic and population growth. During the period from 1947 through 1973 when nominal oil prices were nearly constant, consumption increased at a 7.4% annual growth rate. Then when real prices increased nearly four-fold from 1973 through 1981, consumption growth fell to zero. As real prices declined by 49% from 1981 through 1990, consumption again increased, but at a lower rate (0.8% per year).

Probable reserve additions are subject to a wide estimating error. The U.S. Geological Survey (USGS) has ventured estimates of probable reserve additions for 34 significant oil producing countries of the world, currently accounting for 890 of the 999 billion barrels of world reserves. The USGS projections indicate that world reserves will ultimately be extended by 77%. Thus, the present 1-trillion-barrel world reserve would increase to 1.8 trillion barrels. At present world consumption levels (60 million barrels per day), this reserve is 80 years of supply.

When it becomes clear, perhaps after the turn of the century, that world demand for crude oil will exceed supply at existing prices, then real prices will increase inexorably. If markets are allowed to allocate scarce oil reserves, higher real crude oil prices will ultimately cause probable reserves to expand beyond the USGS estimates. Higher real prices will lead to increased exploration, delay closure of stripper wells, and bring forth production from small new wells and fields that would be uneconomic at prior relatively low prices.

This trend should be reinforced by steady improvements in technologies for exploring for and producing crude oil. Higher oil prices will lead to development of crude oil substitutes and to reductions in the quantity of oil demanded. These supply and demand forces together lead to the conclusion that world oil reserves will be extended into centuries rather than decades.

World crude oil production

World oil production from 1980 through 1990, with projections through 2010, are shown in Table 3-2. A price series (U.S. refiner acquisition cost) is also provided.

The record shows that as the real price of crude declined from 1981 through 1985, world output also declined. However, the entire output reduction was provided by the OPEC producers in a failed attempt to maintain high prices. Non-OPEC producers deemed prices to be attractive and they expanded output. The non-OPEC Centrally Planned Economies (CPE) similarly increased output, probably to the limit of their capability. In 1986, crude prices collapsed as the OPEC output quota regime failed. The major Gulf producers (Saudi Arabia, Iraq, Kuwait, and the United Arab Emirates) increased their production.

The Department of Energy projects a very modest 3.1% increase in World oil production over the next two decades. But OPEC production is expected to increase by 36%, more than offsetting reduced ability to produce by both the non-OPEC market economies and the Centrally Planned Economies. The U.S.

TABLE 3-2   World Oil Production and Price, 1980–1990 with Forecast to 2000.
             (million barrels/day)

Year	OPEC	Non-OPEC Market Economies	Non-OPEC Centrally Planned Economies	Total World	Average Refiner Acquisition Cost (1990 dollars)
Actual					
1980	27.80	21.81	14.56	64.17	45.91
1981	23.76	22.39	14.63	60.78	45.96
1982	20.07	23.42	14.77	58.26	39.99
1983	18.75	24.43	14.92	58.10	34.38
1984	18.81	25.92	14.98	59.71	33.20
1985	17.55	26.80	14.98	59.33	30.73
1986	19.69	26.61	15.44	61.74	16.17
1987	19.85	26.89	15.73	62.47	20.50
1988	21.89	27.06	15.74	64.69	16.07
1989	23.86	26.67	15.43	65.96	18.97
1990	24.96	27.30	15.35	67.61	21.78
Forecast					
1995	26.58	27.35	14.80	68.73	24.00
2000	28.47	26.70	13.60	68.77	25.70
2010	34.05	24.74	10.90	69.69	34.20

SOURCES: Historical record: U.S. Department of Energy (1990a, 1991a). Forecast: U.S. Department of Energy (1990a, 1991b).

Department of Energy (DOE) projects a 57% increase in the real price of crude oil by 2010. Prices substantially lower than the projected $34.20 per barrel would cause non-OPEC output to fall even faster because this production is relatively high cost.

On the demand side, the Middle East is an insignificant consumer of oil and oil products. The industrial nations are the major consumers, but are relatively minor crude oil producers. The 24 Organization for Economic Cooperation and Development (OECD) countries consumed 62.5% of the 1990 world oil production. Five of these countries (U.S., U.K., Norway, Canada, and Australia) produce 95.5% of OECD crude oil. The remaining 19 produce fewer than 120 thousand barrels per day, and seven (Belgium, Finland, Iceland, Ireland, Luxembourg, Portugal, Sweden, and Switzerland) produce no oil at all. (Consumption data are from U.S. Department of Energy, 1991a. Production data are from the Oil and Gas Journal, 1990b.) Among the OECD countries, only Norway and the United Kingdom are self-sufficient Thus, the industrial world is highly dependent on the Middle East for its oil supplies.

Any projection of oil production, consumption, and price must consider the alleged power of OPEC to regulate member output. This is especially true when a 10- or 20-year time period is relevant, rather than a century or more.

The popular press has tended to treat OPEC as a "mighty cartel" with the power to set oil prices in order to maximize OPEC member interests. The long record of price fixing in any industry has shown only mixed success. Success appears to be an inverse function of the number of producers-sellers. When the number exceeds two or three, the probability of success falls sharply. Output coordination and enforcement of controls among the 13 highly independent members of OPEC borders on the impossible. Success is further in doubt given the fact that non-OPEC countries in 1990 accounted for 63% of world output. Only by government intervention on behalf of the members is success assured.

In the United States, market demand prorationing to control oil output was enforced by the federal government from the mid-1930s through 1971. Success was limited in time because not all states consented to be included in the control system. Further, the system was not international and consequently required import quotas beginning in 1959.

There has always been a tendency within OPEC for individual producers to show some semblance of unity at OPEC meeting, but to pursue their individual interests between meetings. The 13 OPEC members include countries of widely diverse critical interests. Some (Ecuador, Algeria, Qatar, and Nigeria) have dwindling oil reserves such that they will be net importing countries within the next five to 30 years, while others have vast oil reserves and resources that lead them to take a long-run view of oil prices and markets. Some (Nigeria and Indonesia) live from hand to mouth while others are among the world's highest per-capita income and wealth nations. Some (Libya, Iraq, and Iran) are strongly anti-American, while others are strongly pro-American and pro-Western in their foreign policy.

Hatred and periodic military conflict between OPEC members (Iran-Iraq, Iraq-Kuwait) make harmonious agreement among members almost impossible. These and other basic interests lead to irreconcilable price-output objectives. Even when meetings produce price-output agreements, there is no effective enforcement mechanism, and economic incentives to agree and then cheat on the agreement are endemic.

The record of cheating within OPEC is found repeatedly. For example, in the first half of 1989 when agreed upon OPEC quotas allowed 18.5 million barrels per day to be produced, actual output exceeded quotas by 12.4%. Table 3-3 shows the record by country. We find that 11 of the 13 member countries were producing in excess of their quotas. The two exceptions, Indonesia and Algeria were probably constrained by reservoir limits rather than quota considerations.

The August 2, 1990 Iraqi invasion of Kuwait introduced new output coordination problems for OPEC. First, Kuwait's production of 1.9 million barrels per day (July 1990) was eliminated. Second, following the successful imposition of U.N. authorized sanctions, oil exports from Iraq amounting to about 3 million barrels per day were eliminated.

Third, other OPEC producers were encouraged to expand production to the limit of their installed ability. Most importantly, Saudi Arabia expanded output from 5.6 million barrels per day in July 1990, to about 8 million barrels per day by spring 1991. As of March 1991, world oil production amounted to 60.19 million barrels per day, compared to 60.23 million barrels per day in July 1990, one month before the invasion of Kuwait. Thus, other nations have expanded

TABLE 3-3   OPEC Quotas and Actual Output by Member Countries, First Half of 1990. (thousand barrels/day)

Country	Quota	Actual	Ratio Quota/Actual
Saudi Arabia	4,524	4,947	1.09
Iran	2,640	2,901	1.10
Iraq	2,640	2,764	1.05
Kuwait	1,037	1,612	1.55
U.A.E.[a]	988	1,604	1.62
Venezuela	1,636	1,667	1.02
Nigeria	1,355	1,492	1.10
Indonesia	1,240	1,200	0.97
Libya	1,037	1,084	1.05
Algeria	695	666	0.96
Quatar	312	388	1.24
Ecuador	230	279	1.21
Gabon	166	195	1.17
Total	18,500	20,802	1.12

SOURCE: Oil and Gas Journal (1989), p. 24.
[a] United Arab Emirates (Abu Dhabi, Dubai, and Sharjah)

output to eliminate the loss of about 5 million barrels per day from Kuwait and Iraq. Fourth, the ability of Iraq to threaten other Persian Gulf producers into restraining their output has been effectively eliminated, meaning that these countries are less constrained in their ability to pursue their own interest independent of either OPEC interests or the interest of individual OPEC powers.

Fifth, as Kuwait and Iraq gradually return to full production in the next year or two, which other OPEC countries will consent to reduce their output and income? The Gulf War has depleted assets of all Gulf producers excepting Iran. Even Saudi Arabia drew heavily on overseas assets to finance the war and will be motivated to pay off debt, rebuild assets, and strengthen their defense capability. Saudi Arabia has announced an intention to expand production to 10 million barrels per day. It is doubtful that Saudi Arabia will be willing to reduce output to accommodate increased production from Kuwait and Iraq, plus higher levels of output from Iran and other Gulf producers. These conflicts between individual country interests and OPEC interests will make output coordination extremely difficult if not impossible. With OPEC production uncertain, forecasts of world oil output for the next two decades must be interpreted with great caution.

World crude oil consumption

Except for changes in inventories, oil consumption is identical to production. The world demand for oil will increase as economic growth occurs, and as population grows. Some nations, including Persian Gulf countries, Mexico, and Venezuela, subsidize oil product consumption by selling gasoline and other products at prices well below world market prices. Other countries, including most of Western Europe, impose taxes on oil products, principally gasoline, that raise consumer prices two- to four-fold above market prices. These subsidies and taxes strongly affect quantities demanded and any changes in such policies would alter demand growth rates.

Concluding comments

When will the world run out of crude oil? From an economic perspective, the answer is "never!" Instead, its real price will rise, leading consumers to conserve its use, and leading to the development of substitutes. Ultimately, perhaps within four or five decades, the world real price of crude oil will have risen substantially and oil will not be used for propelling cars, its major consumptive use today. Rather, automobile producers and consumers alike will find that cheaper and less environmentally damaging energy sources will be developed for automobiles. When the real price in terms of 1990 dollars reaches $100 per barrel, perhaps in about 80 years, then oil products will be reserved for high-value uses such as petrochemical production, and possibly jet fuel for aircraft.

From a physical perspective, oil production is unlikely to decline within our 2010 time horizon. Production will shift from the leading producers of 1990 (the CIS and the United States) and from other producers with declining reserves (the United Kingdom, Algeria, Indonesia), to Persian Gulf producers having both extremely large reserves and unexplored areas.

UNITED STATES CRUDE OIL RESERVES, PRODUCTION AND CONSUMPTION.

U.S. crude oil reserves

The record of United States crude oil reserves is also shown in Table 3-1. The nation's oil reserves reached a peak of 39 billion barrels when the 10-billion-barrel Prudhoe Bay, Alaska field was added to reserves. Since then, reserves have declined persistently to 26 billion barrels as of January 1, 1991. Most of this decline took place during the 1970s, in spite of higher real oil prices during this period. Crude oil price controls were imposed by President Nixon on August 15, 1971 in response to some relatively modest price inflation. They were totally removed as one of the first acts of the Reagan administration in January 1981. However, by this time a "windfall profit tax," better termed an excise tax, was in place, being introduced in the Carter administration as a partial substitute for phasing out crude oil price controls. Although the windfall profit tax legislation is still in effect, the tax produces no significant revenue at today's relatively low crude prices.

Price controls and windfall profit taxes separately and together were negative forces relative to oil exploration and development. Consequently, U.S. crude oil reserves declined by 25% from 1971 through 1982. Since 1982, the decline in U.S. reserves has moderated. Reserves as of January 1, 1991 stood at 26 billion barrels, down 11% from 1982.

Over the entire 44 years from 1948 through 1991 shown in Table 3-1, U.S. reserves have increased 22%. This record further illustrates the point that we cannot forecast production on the basis of current reserves and production rates. In 1947, reserves of 21.5 billion barrels would have indicated only 11 years of U.S. crude supply at the then current rate of production (1.9 billion barrels per year). In fact, between 1947 and 1990, 125 billion barrels have been produced from U.S. reserves and reserves have increased from 21.5 billion to 26.2 billion barrels.

More than 60% of the U.S. original endowment of crude oil has been produced. While the U.S. ranks third among nations in known original endowment, it ranks only seventh today in remaining known reserves (Riva 1991, p. 2).

The U.S. Geological Survey has projected that the estimated January 1, 1991 U.S. reserves, amounting to 26.2 billion barrels, will ultimately be augmented by 71 billion barrels with probable reserve additions (Masters, Root, and Attanasi, 1990). Exploration for and production from these projected reserves must pass not only the geological and economic tests, but also political tests. At present, environmental organizations seek to block even leasing from the most promising oil provinces—Alaska's National Wildlife Refuge and the Santa Maria Basin offshore from California.

U.S. crude oil production, consumption, and imports

While U.S. oil reserves are only 2.6% of world reserves, U. S. production currently at 7.6 million barrels per day accounts for 10.9% of world production. With a reserve-to-production ratio of 9.4, paired with the relatively high rate of historical exploration in the U.S., the nation's current production rate is not sustainable.

The U.S. Department of Energy has prepared a forecast by 5-year intervals from 1990 through the year 2010. The "reference case" under this forecast assumes that crude oil prices (1990 constant dollars) will rise from $22 per barrel in 1990, to $34 per barrel in 2010. Even with this economic stimulation, U.S. crude production is expected to decline steadily from 7.6 million barrels per day in 1990, to 4.4 million barrels per day in 2010 (U.S. Department of Energy, 1991b, p. 52). Counting natural gas liquids, refinery processing gains, and synthetic petroleum production, the total U.S. petroleum supply is expected to decline 28% from 9.9 million barrels per day to 7.1 million barrels per day by 2010.

Private forecasting firms offer generally more optimistic expectations. While the Wharton group (WEFA) agrees almost exactly with the DOE 2010 forecast for crude oil production, McGraw Hill's Data Resources, Inc. (DRI) expects crude production to be at 4.7 million barrels per day in 2010, and the Gas Research Institute (GRI) forecasts an optimistic 6.4 million barrels per day production level (*Ibid.*, p. 38). These forecasts are shown graphically in Figure 3-3.

With U.S. petroleum consumption expected to increase 21% over the next 20 years, imports of crude oil plus refined products are projected to increase 71% to 21.1 million barrels per day by 2010. Thus, oil self-sufficiency for the U.S. is likely to decline from 58% in 1990 to 36% in 2010.

Historic political responses to the problems posed by increasing oil supply dependency have been short-run fixes that exacerbated the problem in the long run. Federal tax subsidies for oil producers (percentage depletion allowance and expensing of intangible drilling costs) have stimulated early oil production at the expense of future production. Oil import quotas raised domestic crude prices to artificially high levels thereby artificially stimulating domestic production, again sacrificing future supply security. The present system of Strategic Petroleum Reserves discussed in another chapter offers a solution wherein the nation may import relatively low cost crude from where it is abundant, and achieve supply security via stored reserves.

Crude oil prices and costs

The record of real crude oil prices for the United States is also shown in Table 3-2. This record may mislead because oil prices were hitting all-time high levels in 1980–81. A longer time span is shown in Figure 3-4, with prices shown in both real and nominal terms. We find that real prices were nearly constant from 1968 until ownership problems surfaced in the Middle East in the early 1970s. With massive nationalization in 1972 and the Israeli-Egyptian War in 1973, crude prices doubled. Then coinciding with the 1978 Iranian Revolution and a sharp decline in crude production from that country, prices again increased nearly two-fold. The 1980 invasion of Iran by Iraq again sent prices higher as production from both countries declined. Average refiner acquisition cost of imported oil reached a peak of $39 per barrel in February 1981.

From 1981 to date, oil prices have been weak, interrupted only by the Persian Gulf crisis of late 1990. Attempts by OPEC to restrict output and stabilize crude prices were effective only in preventing even more severe price erosion. As of June 1991, prices were relatively stable at about $20 per barrel. In 1968

Production (bbl./day in millions)

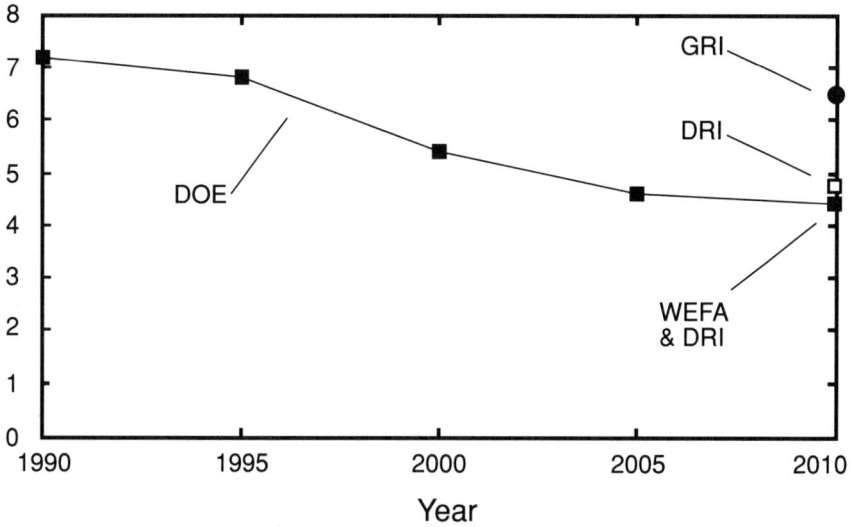

FIGURE 3-3   Actual 1990 and Forecast U.S. Crude Oil Production.   SOURCE:
U.S. Department of Energy (1991b), pp. 38, 52. DOE = Department of Energy; DRI = Data
Resources, Inc.; GRI = Gas Research Institute; WEFA = Wharton Economic Forecast
Associates.

Price ($/bbl.)

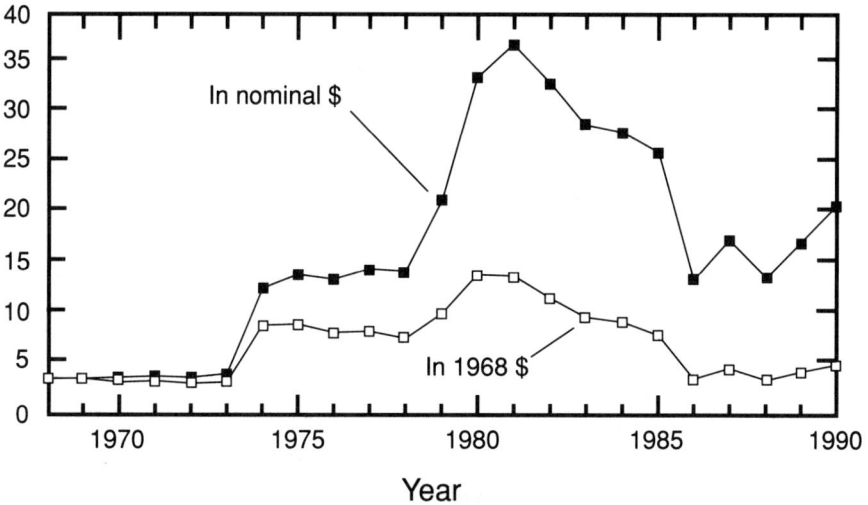

FIGURE 3-4   Nominal and Deflated Crude Oil Prices.

dollars, real prices increased from $2.90 per barrel to $6.37 per barrel in 1991. Thus, real prices have increased at a 3.5% compound annual rate over 23 years. Table 3-2 shows the Department of Energy expectation that real prices will continue to increase from 1991 through the year 2010 at compound annual rate of 2.9%. There is a tendency in the DOE forecast for prices to rise more in the decade after 2000 than before. The analysis provided above relative to world oil reserves and production indicated that price increases after the turn of the century are likely to be substantially higher than in the present decade.

If this modest real price increase comes to pass, then even without new taxes on oil and oil products, conservation should be expected from normal market incentives. Furthermore, oil substitutes would receive market encouragement.

The wellhead cost of producing crude oil is not a single value. Rather, cost per barrel on a well basis is a function primarily of output per day. The large Saudi Arabian wells producing over 10,000 barrels per day have production costs less than $1 per barrel. For stripper wells (wells producing less than 10 barrels per day) production ceases when incremental costs exceed the wellhead value of the crude. Therefore, the highest cost wells in the U.S. have costs slightly less than about $20 per barrel market value, minus the cost of transporting the oil from wellhead to market. Thus, costs range from about $1 per barrel for the most prolific wells, to current wellhead value.

Potential crude oil substitutes

If world and U.S. oil consumption is to decline in the future, a decline must occur in the transportation sector. Data for the United States show that as of February 1991, 62% of the oil used was in transportation. Furthermore, petroleum products accounted for 97% of the primary energy consumed in transportation. Substitutes for oil in transportation will occur as oil prices rise and stimulate research and development of substitute technologies.

Potential crude oil substitutes may be classified as direct substitutes, or indirect substitutes. (For a cost review of the major alternative fuel technologies and electric power technologies, see Mead and Denning, 1991.) The direct substitutes consist of new technologies for producing synthetic oils that are so close to natural that they may be refined into the conventional petroleum products including gasoline, jet fuel, diesel fuel, etc. There are two such substitutes available to the U.S.—synthetic oil from oil shale or from coal.

The United States is generously endowed with both oil shale and coal resources. World oil shale resources have been estimated at 88 trillion tons, which may be converted into about 2 trillion barrels of oil (World Energy Conference, 1983, p. 51). The United States has about one-third of this world reserve, or about 650 billion barrels. This is almost exactly equal to the entire Middle East crude oil reserve. But, alas, the cost of converting this resource into synthetic crude oil is estimated in the range of $45 to $60 per barrel in 1987 dollars (Mead and Denning, 1991, p. 248). With crude oil available at $20 per barrel, the oil shale resource is not a reserve. It will be uneconomic to recover and process until the real price of crude oil increases nearly three-fold.

The most recent evidence of infeasibility for oil shale is that, effective June 1, 1991, Union Oil Co. of California (Unocal) closed the only remaining shale oil production plant in the U.S. This plant, which started operations in August 1983, reached a production peak of 7,000 barrels per day in 1989. It existed only with a subsidy from the federal government that guaranteed a selling price that amounted to $49.77 per barrel in 1990. According to Unocal management, that oil had a market value of $24.17 per barrel (Los Angeles Times, 1991a). The company lost money during each year of operation and argued that a selling price of $50 per barrel would be necessary for profitable operation.

Regarding development of world oil shale resources, one observer wrote in 1983 as follows (Sande, 1983):

In 1980, the new decade appeared to herald unprecedented development of [oil shale and bituminous sands]. Forecasters were predicting 5 million barrels per day of synthetic oil production by the year 2000; work was underway on a broad technical front in all geographical regions. Today [1983] it is unlikely that anyone is considering construction of a world-scale project in this decade. (p. 48)

In Venezuela, the Orinoco project, planned to upgrade 26,000 cubic meters per day of heavy crude has been shelved. . . . (p. 48)

In 1983, the U.S. price of imported oil averaged $29 per barrel, having fallen from $34 per barrel in 1980. The 1991 price of about $20 per barrel makes development of synthetics even less viable.

Similarly, world coal reserves are large, and coal can be converted into synthetic crude oil. World recoverable reserves of anthracite and bituminous coal are estimated to be 493 billion short tons, or 2,219 billion barrels of oil equivalent (World Energy Conference, 1983). The United States has about

34% of this reserve, or 760 billion barrels of oil equivalent. But the cost of coal liquefaction is prohibitive at present. The proposed Breckenridge coal liquefaction plant, designed to produce 4.2 million barrels per day of synthetic oil per year, involved a conversion cost of $96 per barrel in 1987 dollars (Mead and Denning, 1991, p. 248). This is nearly five times the value of the oil product.

There are several promising alternative indirect substitutes for crude oil. Any electric power alternative is an indirect substitute, in that electric power may be generated, stored in batteries, and used for transportation in lieu of gasoline, which comes from crude oil. New technologies for lower cost electric power generation using nuclear energy are now being developed by Westinghouse, General Electric and others. These technologies involve smaller scale (about 600 megawatts instead of about 1200 megawatts), more prefabrication of modules, and increased safety features. If the licensing and certification procedures can be shortened and litigation reduced, then costs for nuclear power can be significantly reduced. Nuclear power plants operate around the clock. Electric power demand follows a daily cycle in which usage is very low between midnight and about 6:00 a.m. The economic consequence is that some electricity generated at night has no market and its economic cost is near zero. Batteries for cars and light trucks can be recharged in the off-peak hours at very low economic cost.

Electric vehicles have two advantages—their use reduces our dependency on foreign crudes; and even with the current mix of electric power generating technologies, pollution emissions in transportation would be reduced. For example, grams per mile of pollution from carbon monoxide would be reduced 50%, nitrogen oxides by 33%, volatile organic compounds by 99%, and carbon dioxide by 54%. (These estimates are for use in the Chrysler TEVan. See Electric Power Research Institute, 1991.) With nuclear power rather than the present mix of hydro, oil, natural gas, nuclear and geothermal, these pollutants fall to approximately zero, in exchange for a very low risk of a catastrophic accident and potential problems arising out of nuclear waste storage.

There are also two significant disadvantages of electric vehicles. First, the initial capital cost is high. Under present technology, the General Motors electric van initial cost is $50,000. This includes $7,000 to $8,000 for a lead-acid battery pack and is about three times the cost of a conventional van. The battery pack must be replaced every 30,000 miles (*Ibid.*, p. 12). Second, the range from one battery charge is limited. The Ford modified Escort is expected to have a

100-mile range. The hybrid modification of the Escort in which a 10-horse-power gasoline motor is added is intended to increase the range to a maximum of 250 miles, at the cost of a small addition to air pollution and some additional capital cost. In addition, methanol, ethanol, and natural gas are oil substitutes for transportation, each with unique problems including economic viability.

As indicated above, real crude oil prices are expected to rise after the turn of the century, when evidence appears that the ability to produce crude oil declines relative to demand at prevailing prices. At the same time, the real cost of crude oil substitutes is likely to decline with technological advances. These substitutes (backstop technologies) will come on-line when they become economically viable relative to crude oil. Government subsidies may accelerate the process of product introduction, but they do not reduce the economic costs.

Concluding comments

Crude oil reserves in the U.S. have been declining since 1971 when the 10-billion-barrel Prudhoe Bay reserve was added. Given the fact that the U.S. is the most explored nation in the world, discovery of a new super-giant, or even giant oil field in the U.S. has a very low probability. The best prospects are in the Alaska National Wildlife Refuge and the Santa Maria Basin offshore from California. However, exploration and production in these two areas are strongly opposed by politically powerful environmental organizations.

Crude oil production in the U.S. is declining, as it must with falling reserves. The Department of Energy forecasts that current production at 7.6 million barrels per day will decline to 4.4 million barrels per day in the two decades ending in 2010. The U.S. will fill the gap by importing more crude from the Middle East where reserves are concentrated. As a result, U.S. crude oil self-sufficiency will decline from the present 58% to 36% by 2010. Crude oil prices, adjusted for inflation, will rise, particularly after the turn of the century. The Energy Department forecasts that crude prices will rise from the present $20 per barrel in 1990, to $34 per barrel in 2010. Higher prices will lead to more conservation and to the development of substitutes for oil.

WEST COAST AND CALIFORNIA CRUDE OIL RESERVES, SUPPLY, AND CONSUMPTION

To understand the probable future supply of crude oil available to the West Coast and California, we will identify the major sources of crude oil.

To begin with, Washington and Oregon have no significant oil reserves or production. California has been a major oil-producing state for most of the present century. However, conventional (light) crude oil production from California reached a peak in 1954 and has been in a persistent decline thereafter. Nonconventional (heavy oil) sources have been expanding over this same time period with the result that total California onshore production has been relatively steady for nearly four decades.

The only nonconventional oil in California is called "heavy crude." It is found relatively near the surface and is concentrated in Kern County. Heavy oil is generally defined as crude oil with a gravity rating less than 20° API. It tends to be relatively high in sulfur content and contains other impurities. Consequently, this tar-like crude is more expensive to refine and converts into more low-value residual fuel oil and less high-value gasoline. Further, its production requires steam injection into the reservoir, a process that creates some air quality problems.

Also offsetting the loss of conventional production is discovery of large reserves in the marine environment. The Santa Maria Basin, along the coast north of the Santa Barbara Channel as far as the Pismo Beach area, has been described as the most promising oil province remaining in the lower 48 states. However, its development, production and transportation have been delayed by environmental objections and litigation. Further, like the Kern County crude, much of the Santa Maria Basin crude is of low quality due to low gravity and high sulfur as well as other impurities. If the political problems arising out of environmental concerns can be resolved, then crude from this source may substantially delay reliance on imports for California.

Another major source of oil for the California market is shipments by tanker from Alaska's North Slope, drawing on the huge Prudhoe Bay field. However, this field, like California conventional onshore, is also in a declining phase. Markets in Alaska and the Pacific Northwest have a prior economic claim on this oil because the transportation costs are lower than for California delivery. Profit-maximizing producers would be inclined to deliver to the nearest markets because the lower transport costs would yield a higher net-back to wellhead value. Both markets are expanding. Consequently, with the passage of time less oil from the Alaska source will be available to California.

Imports from Indonesia and the Middle East are available to the California market. As U.S. sources decline, imports will rise to fill the gap. At present, imports from Indonesia are limited to certain specialty oils such as high-gravity oil for blending. Imports into the West Coast from the Middle East are insignificant. Our study will survey all the supply sources listed above year-by-year through 2000.

The California demand for crude oil must also be estimated. The quantity demanded depends on such economic factors as the price of oil and its products, population growth, income of the people in the State, and the cost and availability of substitutes for oil products. Demand functions will be estimated drawing on past relationships between the important causal variables and the quantity demanded.

Finally, with supply sources identified by year and information in hand showing reliance on foreign sources, we will explore the need, if any, for public policies to mitigate adverse economic impacts of a supply disruption.

Why is crude oil supply and demand important for California and the West Coast?

Consumers in California, Oregon and Washington have benefitted from the "West Coast oil glut." Except for minor imports of unique crudes, the West Coast has received all of its crude supplies from in-state and Alaskan sources. Since 1977 when crude oil began flowing through the Alyeska oil pipeline from Prudhoe Bay, the supply of crude oil on the West Coast has exceeded demand at prevailing prices with the excess being shipped to the Gulf of Mexico area, the Caribbean, and the East Coast via Panama. (Small quantities of crude have been shipped east through the Four Corners Pipeline.) Exports have been prohibited by act of Congress. Consequently, the cost of crude oil on the West Coast has been below Gulf of Mexico costs by an amount approximately equal to the cost of shipping crude from the West Coast to the Gulf region. This cost difference varies from marginal cost only, to full cost, depending on the condition of the tanker market. The cost difference has been $2 per barrel, plus or minus about $1 per barrel.

Since 1977, California crude oil demand has been increasing, and a steady decline rate has begun for the Prudhoe Bay field, with the result that the West Coast oil glut will become a deficit in the mid-1990s. This deficit will be filled with increased imports placing California under some risk due to a possible disruption in foreign supplies. When imports begin, crude oil prices on the West Coast will rise to approximate parity with Gulf of Mexico delivered crude import prices.

The marginal sources of crude for both markets will be the Middle East. Cost of marine transport from the Middle East to the two markets is approximately the same. Consequently, the delivered price will be approximately the same. When this price parity occurs, the West Coast price will increase by about $2 per barrel (10%) and the West Coast price advantage will disappear. Petroleum product costs to consumers on the West Coast will rise in proportion. Therefore, two answers come to the fore regarding the question—why is crude oil supply and demand important for the West Coast? They are, first, the cost of crude oil on the West Coast will increase approximately 10% relative to other U.S. and world markets, and second, the supply security provided by the West Coast oil glut will disappear and the West Coast, like the rest of the country, will be subject to potential supply disruptions.

West Coast crude oil reserves

Reserves of crude oil in California including the California Outer Continental Shelf (OCS) are estimated at 5.8 billion barrels (U.S. Department of Energy, 1990b, pp. 73, 82). This is 22% of the 26.2 billion barrels total U.S. reserves (Oil and Gas Journal, 1990a, p. 45). Onshore reserves, estimated at 4.8 billion barrels, account for 83% of the total state reserves, with the remaining 1.0 billion barrels found in the federal OCS.

Annual additions to reserves in California have usually exceeded annual production. Consequently, reserves have increased from 3.7 billion barrels in 1950 to 5.8 billion barrels in 1989. The record of California crude oil reserves is shown in Table 3-4. The stimulating effect of much higher real prices in the 1970s is apparent in the sharp increases in crude oil reserves shown for 1980 and 1985.

Onshore conventional California oil prospects have been extensively explored during the current century. Discovery of new onshore reserves is very unlikely and should not be expected. Any increase in reserves will be limited to

TABLE 3-4   California Crude Oil Reserves, including Heavy Oil and Outer Continental Shelf Reserves. (million barrels)

Year	Reserves
Actual	
1950	3,733
1955	3,801
1960	3,658
1965	4,567
1970	3,984
1975	3,647
1980	5,470
1985	5,801
1989	5,803
Forecast	
1994	4,719
2000	5,215

SOURCES: 1950-1989: U.S. Department of Energy (1990b); 1994 and 2000: Ibegbulam (1991).

the California heavy oils that will require higher real prices to convert resources to reserves, and to offshore resources where leasing, exploration and development are currently on hold due to environmental politics.

Future California reserves have been estimated by Ibegbulam (1991). Reserves in each year are modeled as a function of prior year reserves and current real crude oil prices. This forecast concludes that California reserves will decline modestly through 1994 as real crude prices decline, and increase thereafter through 2000 to a level approximating current reserves (*Ibid.*, p. 221). The incentive for the increase is higher real prices expected in the last half of the 1990s.

West Coast crude oil supply and demand

The California Energy Commission (CEC) has forecast crude oil supplies available from onshore and offshore California, Alaska, and foreign imports through the year 2009. This forecast was made in 1989, prior to the Gulf War and was based on crude price assumptions that are well below actual prices. The CEC supply forecast is shown in Table 3-5 along with the UCSB forecast.

In addition to reserves, demand and supply functions have been estimated separately at UCSB by Ibegbulam. Drawing on that study, we will briefly out-

TABLE 3-5   Supply and Demand Forecasts for Onshore and Offshore Crude Oil
             Production, State of California. (million barrels/year)

Year	Supply forecast CEC[a]	UCSB[b]	Demand	Surplus shipped to Gulf of Mexico[b]	Deficit filled by imports[b]
1990	815	825	618	207	—
1991	819	781	627	154	—
1992	771	713	636	77	—
1993	720	652	644	8	—
1994	694	607	650	—	43
1995	618	540	650	—	110
1996	546	480	650	—	170
1997	492	436	651	—	215
1998	454	407	652	—	245
1999	419	379	654	—	275
2000	413	382	655	—	273

[a] California Energy Commission (1989).
[b] Ibegbulam (1991), p. 235.

line the model and utilize its conclusions. Crude oil production from onshore
and offshore California in year t is specified as a function of total reserves in
year t-1. Total reserves are a function of prior year reserves and crude oil prices,
the latter being a proxy for costs of exploration and development. Crude oil
price was entered endogenously using the UCLA forecast of crude oil prices
for the U.S. West Coast prices are shown to be about $2 per barrel below U.S.
prices until 1994, when what has been called the "West Coast discount" disap-
pears.

Like the supply function, the demand for crude has been estimated using
current econometric techniques. Historical relationships have been drawn upon
extending back to 1960. The demand for crude oil is derived from the demand
for all of the products into which crude is converted. Therefore, to estimate the
demand for crude, demand functions for each of its final products were estimated, then converted into "crude oil equivalent" terms. Demand has been
estimated separately for each of the following oil products: gasoline, distillate,
residual oil, jet fuel, aviation gasoline, kerosene, lubricants, asphalt and road
oil, and "other."

The model is dynamic, making use of a one-period lag on the dependent
variable in the estimating equation. Demand is regressed on its price, income or
a measure of output, a variable specific to the oil product (if any is available),

and the one-period lag of the dependent variable. Each equation is expressed in double logarithmic form. This formulation produces coefficients that are short-run elasticities. Long-run elasticities are obtained by dividing the short-run elasticities by one minus the coefficient of the lagged dependent variable. The results produce coefficient signs that conform reasonably with expectations and are generally significant. The elasticities appear to be reasonable. These supply and demand forecasts are shown in Table 3-5.

The forecast results indicate that the current West Coast oil glut amounting to about 207 million barrels per year being shipped to the Gulf of Mexico area, will disappear in 1994 when a deficit amounting to 43 million barrels for the year will appear. Thereafter, imports will increase steadily to about 275 million barrels per year by 2000.

The major factor leading to this shift from a surplus to a deficit area is declining supplies, rather than increasing consumption. The latter increases modestly, consistent with increasing population and income, partially offset by continued conservation incentives driven by an upward trend in real prices for crude oil.

Crude supplies from Alaska. The giant Prudhoe Bay oil field on Alaska's North Slope is now in a decline phase. After peaking at a rate of 1.6 million barrels per day in 1987, the forecast decline rate to the year 2000 is 6.9% annually, at which time output is expected to be down to 632 thousand barrels per day (Alaska Department of Natural Resources, 1991). The historical record and projections are shown in Figure 3-5.

Other Alaskan fields produced an additional 400 thousand barrels per day in 1987, so that total Alaskan production was 2 million barrels per day. Some of these smaller fields are still expanding output, while some are declining, but at a slower rate. In addition, Cook Inlet production, which stood at only 45 thousand barrels per day in 1987, is expected to be shut down in 1993. For Alaska in total, the 1990 to 2000 decline rate is expected to be 5.9% annually. After that, the decline rate for the next decade is projected at 11.9%.

Deliveries of Alaska crude. In 1990, the State of Alaska produced 2,007 thousand barrels per day of crude oil, condensate, and natural gas liquids, of which about 264 thousand barrels per day were used in Alaskan refineries. This left 1,743 thousand barrels per day available for delivery to other U.S. destina-

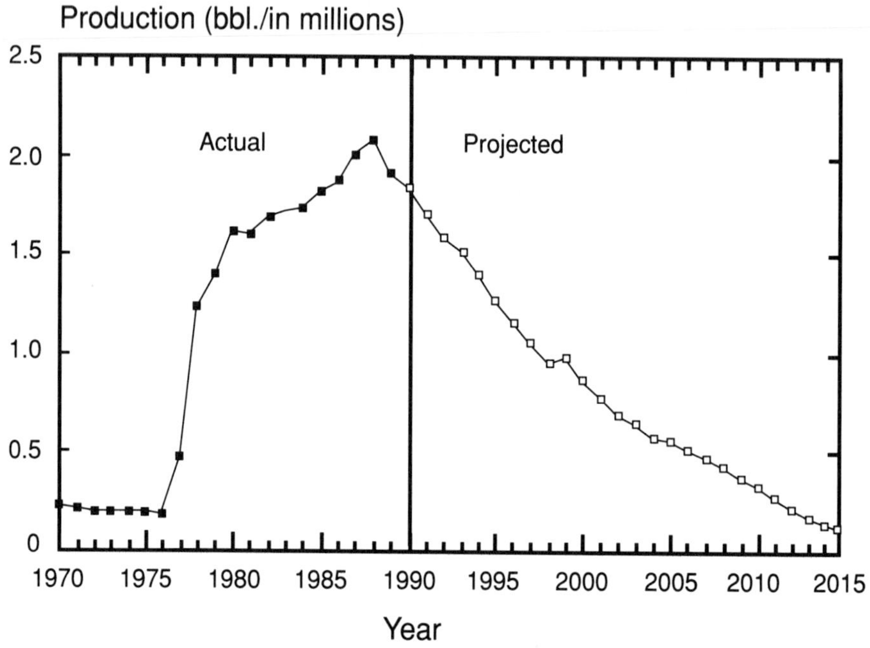

Production (bbl./in millions)

FIGURE 3-5   Alaska Crude Oil Production. SOURCE: Alaska Department of Natural Resources (1991), pp. 4–11.

tions. Figure 3-6 graphically shows the pattern of deliveries. Details are provided in Table 3-6.

Imports augmented the West Coast supply by 267 thousand barrels per day. Hawaii received 130 thousand barrels per day. About 523 thousand barrels per day were dropped off in Washington and Oregon, with 97% of it going to the refineries around Puget Sound in Washington. This left 1,357 thousand barrels per day available for California. The California market required about 800 thousand barrels per day to meet its refinery needs. The remaining 557 thousand barrels per day, termed the West Coast surplus, was shipped via Panama to refineries near the Gulf Coast and the mid-continent, to the Caribbean and to the East Coast.

The East Coast market is the least profitable market for Alaskan oil, with other destination points utilizing the Panama pipeline in the next least profitable position. Shipments to these markets would not occur except for the ban on exports from Alaska's North Slope (ANS). Instead, energy and other resour-

TABLE 3-6   Distribution of Alaskan Crude Oil, 1990. (thousand barrels/day)

Source/Destination	Amount
Oil shipments from Port of Valdez	1,743
Plus imports into PAD V	267
Less deliveries to Hawaii	-130
Less deliveries in Puget Sound, Washington	-508
Less deliveries in Oregon	- 15
Equals crude available to California	1,357
Less deliveries in California	-800
Equals West Coast surplus	557
Deliveries via Panama to the Gulf Coast, the Virgin Islands, and the East Coast	-557
Balance	0

SOURCES: U.S. General Accounting Office (1990); Alaska Department of Natural Resources (1991).

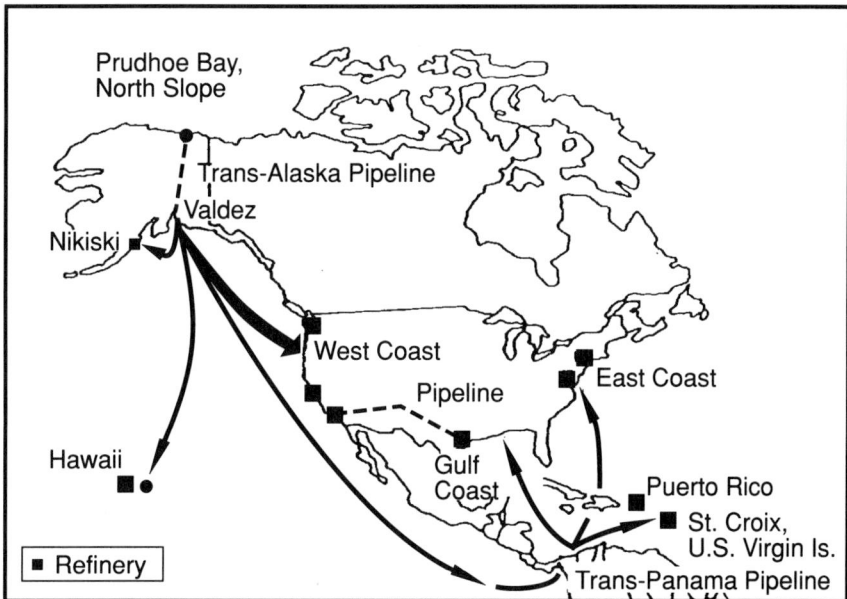

FIGURE 3-6   The Pattern of Alaska Crude Oil Deliveries under the Export Ban.
SOURCE: U.S. General Accounting Office (1990), p. 4.

ces would be saved by shipping some ANS oil to the Orient and replacing these exports with like amounts of imports from Mexico, Venezuela and the Middle East.

California conventional (light) crude production. Crude oil may be divided into two classes according to its gravity rating. Conventional (light) crudes having API gravity ratings above 20°, and heavy oil with gravity ratings of less than 20°.

Conventional crude oil production from California reached a 1954 peak with output at 733 thousand barrels per day. A persistent decline dropped output to 292 thousand barrels per day in 1990. The observed increase in output beginning in 1977 and lasting through 1982 is accounted for primarily by the U.S. Department of Energy bringing the Elk Hills Naval Petroleum Reserve into production. The probability is high that conventional crude output from California wells including Elk Hills will continue to decline into the foreseeable future.

Production from the Elk Hills field. In 1909 President Taft, by executive order, withdrew 38,073 acres of potential oil land from the public domain located in the Elk Hills oil province. With the later addition of 8,022 acres, the Elk Hills Naval Petroleum Reserve was established. This reserve, 81% owned by the federal government and 19% by Chevron, was partially developed but remained in shut-in status until 1976. The Reserve was estimated to contain 1.36 billion barrels of oil as of 1976. By 1990, reserves had declined 62% to 0.52 billion barrels (Hogan, 1990).

Prior to 1976, annual production was limited to output from testing. During the Arab-Israeli war of 1973 and its consequent oil embargo, use of the Elk Hills Reserve was prevented by jurisdictional quarrels in Congress. With these matters resolved, Elk Hills was placed in production status in 1976 with volume production occurring in 1977. Beginning about 1977, the field was producing at its Maximum Efficient Rate (MER).

Unfortunately, high-level production corresponded with the beginning of production from Prudhoe Bay and the consequent "West Coast oil glut." Elk Hills production reached a peak in 1981 at 173 thousand barrels per day and declined by more than half to 81 thousand barrels per day in 1990. The production record is shown in Figure 3-7. Instead of contributing to an oversupplied California market, the Elk Hills field might have been used for a petroleum reserve as originally intended and supplementing the Strategic Petroleum Reserve. (For an elaboration of this reserve proposal see Mead and Sorensen, 1971.)

Planned production through the year 2000, when output will be about 48 thousand barrels per day, is also shown in Figure 3-7. Thus, production at the reservoir MER from 1977 to date has added to the West Coast oil surplus problem. Then beginning in 1994, when that surplus is expected to disappear, declining production from Elk Hills will contribute to the expected accelerating deficit.

Heavy oil production from Kern County. Heavy oil is a tar-like hydrocarbon that ordinarily requires injection of heat into the reservoir to stimulate production. For this purpose, steam is generated using either oil or natural gas as the fuel source. In the early days of heavy oil production in California, about one-third of the crude oil output was burned in boilers to generate the required steam. Later, residual fuel oil was purchased and burned. Recently, producers

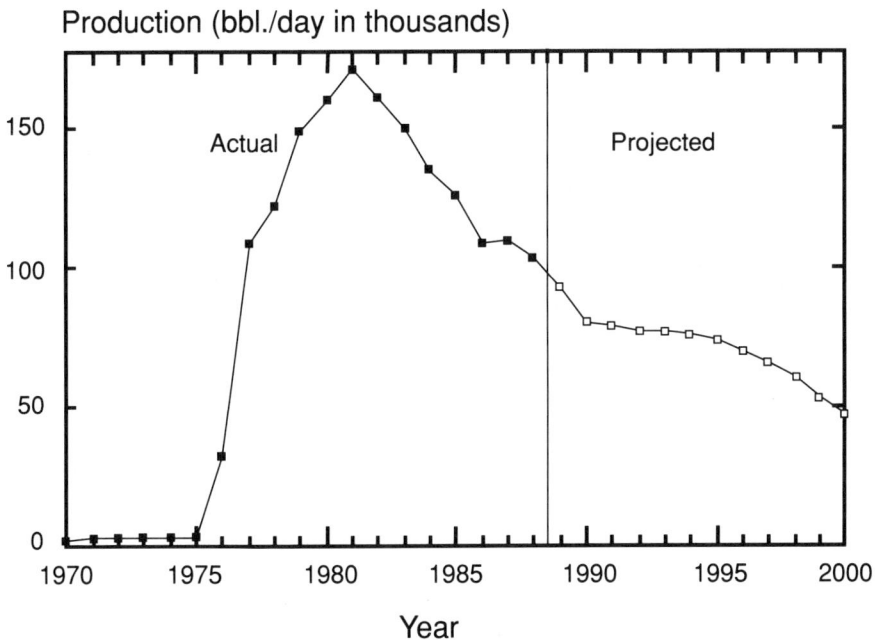

FIGURE 3-7   Crude Oil Production from Elk Hills Field.

have been shifting to natural gas in order to minimize air pollution. Cogeneration has also been introduced with electricity being generated in steam turbines and the residual heat injected into the reservoirs.

Our forecast of heavy oil production is based on in-depth interviews with the principal producers of heavy crude oil. The critical research question concerns how producers will react to lower and higher crude oil prices. At what price will existing production wells be shut in if prices fall, and what price will bring forth new capital investment to expand production from known heavy oil reserves?

As of December 31, 1987, reserves of heavy oil in California were estimated at 5.2 billion barrels. This amounts to 21% of U.S. crude reserves and at least 70% of California onshore reserves. About 95% of U.S. heavy oil reserves are located in California. The small remainder is found in Texas.

Current estimates of California heavy oil reserves probably understate the true reserves for three reasons. (1) Some fields that are classified as containing heavy oil have no reserve estimates. (2) Heavy oil production is still relatively young and some fields that will be found and produced in the future have not yet been identified. (3) Oil is a nonrenewable resource and its price is quite likely to increase sharply in the next century as world oil reserves decline and demand continues to expand. Reserves are a function of product prices and cost of production. Real prices should move upward into the distant future and technology in this new industry should continue to improve, lowering production costs. For reasons of this kind, Guerard speculated that "it is reasonable to expect an eventual recovery of at least 9 billion barrels" (Guerard, 1984).

Heavy oil has become a major source of California's crude production. In 1945, heavy oil represented 25% of the state's production. By January 1989, this share had increased to 70%. Light oil output peaked in 1954 at 720 thousand barrels per day and declined 39% to 298 thousand barrels per day by 1989. Heavy oil production has followed an opposite path, increasing persistently since 1941 from 120 thousand barrels per day to 700 thousand barrels per day in 1989. This record is shown graphically in Figure 3-8. The important role of heavy oil in California production is clearly illustrated.

As of June 14, 1991, the wellhead value of 13° gravity heavy Kern River oil was $12.00 per barrel. In contrast, light oil (Belridge 31° gravity) was valued at $15.90 per barrel, a spread of $3.90 per barrel in the posted price (Chevron, 1991). Our interviews with producers indicated that variable costs of production would be covered with heavy oil prices as low as the $6 to $8 range. This range becomes a close-down area. New field development requires coverage of all costs including a normal rate of return on the investment. Producers indicated that fields of a quality similar to those in current production would require an expected future price of $13 per barrel (Mead, 1989).

Our survey concludes that heavy oil production will increase from 772 thousand barrels per day in 1990, to a peak of 865 thousand barrels per day in 1995 and will decline sharply thereafter to 642 thousand barrels per day in the year 2000. At its 1995 peak, output will amount to about 78% of California oil production.

California production excluding Elk Hills and the heavy oils. Output of conventional onshore and offshore crude oil excluding Elk Hills is shown in Table 3-7. This record is of interest because it is the historical base of California oil production. Elk Hills production is under federal government direction and is managed politically. Heavy oil, as indicated above, is still expanding and is highly responsive to heavy crude oil prices. By excluding heavy oil and Elk Hills, we can focus on the production trend of the historical base.

The evidence presented in Table 3-7 shows a rapid decline in production. That is, production from existing wells is declining much faster than reserves are being expanded by new discoveries and extensions of known fields. Table 3-7 shows that production from this historical base has declined at a 4.4% annual rate. Reference again to Figure 3-8, showing a decline in light-oil production, obscures the depletion problem. The decline rate shown there for light oil including Elk Hills is 2.9%. Thus, the loss of production from the historical base of California oil production is clear and persistent. In the absence of any forecast of production from this historical base, the best guess is that the observed 4.4% decline rate will continue through the turn of the century.

Concluding comments. Conventional crude oil production in California peaked out in 1954. Since then, heavy oil production has been expanding to offset that decline with the result that total California production has been nearly constant through 1990. However, heavy oil production is forecast to reach its maximum level in 1995, and then to enter a decline phase. Production from the important Elk Hills field is already declining. Known reserves are available from the California OCS, but production is blocked by environmental politics

TABLE 3-7   Conventional California Crude Oil Production excluding Elk Hills,
             1970–1990. (barrels/day)

Year	Production
1970	521,884
1971	494,413
1972	450,777
1973	425,845
1974	403,586
1975	372,028
1976	328,332
1977	317,449
1978	303,848
1979	289,777
1980	276,737
1981	276,058
1982	278,979
1983	229,985
1984	244,318
1985	234,381
1986	242,622
1987	200,452
1988	215,639
1989	204,445
1990	212,756

SOURCES: Conservation Committee of California Oil Producers (1990); Hogan (1990).

with difficult-to-predict outcomes. Similarly, new OCS leasing and exploration
is being held up on environmental issues.

An important source of oil to the West Coast and California is production
from Alaska's North Slope. However, the Prudhoe Bay field is now in a decline
phase and production from all Alaska sources is forecast to continue declining
at least through the year 2010.

The foregoing conclusions concern supply; demand on the West Coast is
forecast to expand at a modest pace. Crude prices (inflation adjusted) are forecast to rise beginning in the mid-1990s. This price increase will lead to additional conservation efforts. Nevertheless, crude oil markets will be subject to a
"double whammy" effect through both supply and demand.

An important crossover point is forecast for the year 1994 when the West
Coast surplus, now being shipped to the Gulf of Mexico and points east, will

Production (bbl./day in thousands)

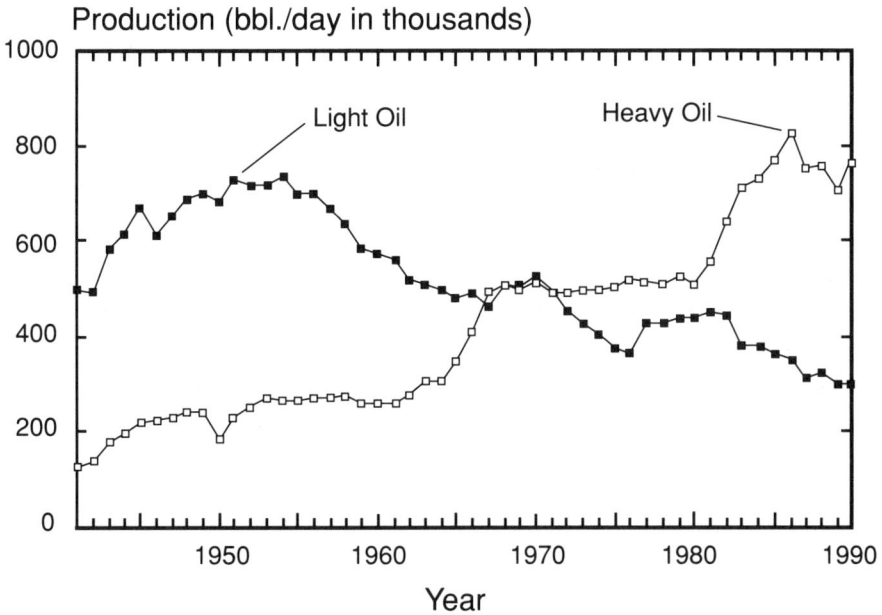

FIGURE 3-8   Calfironia Light and Heavy Crude Oil Production.

become a deficit. In 1994, imports will begin increasing, rising steadily about 273 million barrels per year by the year 2000.

Two important consequences follow. First, the price of domestic crude oil will rise about $2 per barrel (about 10%) to equal that of imported crude in the Gulf of Mexico and East Coast. Second, as imports increase after 1994, the West Coast will join the rest of the country in being dependent on uncertain imports. This dependency will increase for the entire nation as the United States' ability to produce crude oil declines.

SUMMARY OF CONCLUSIONS

The world will never run out of oil. As world oil reserves begin to decline sometime after the turn of the century, real crude prices must be expected to rise. Thus, future generations will have crude oil, but it will be increasingly expensive. Higher real prices will lead to additional conservation efforts and worldwide oil consumption is expected to begin a long decline. Centers of oil production will inevitably shift to the Middle East from today's leading producers—the Commonwealth of Independent States, the United States, and the United Kingdom.

Ultimate higher prices for oil lead not only to conservation in its use, but to the development of substitutes. The major use of crude oil worldwide is for automobile transportation. As the price of gasoline rises, electric cars will become economically viable. The environment (sharply reduced air pollution, acid rain, and greenhouse effects) will benefit, especially if future power generation is from nonpolluting nuclear sources instead of coal, oil, and to a lesser degree, natural gas. These beneficial environmental consequences require solution to the largely political problems of nuclear waste storage.

The United States and the CIS, as the two largest oil producers in the world (prior to the Persian Gulf War), are producing oil at unsustainable rates and both are in a decline trend. Oil production in the United States reached a peak in 1970 when production for the year averaged 9.6 million barrels per day. By spring 1991, production had fallen to below 7 million barrels per day. Production is forecast to decline further to 4.4 million barrels per day in 2010. At that time, dependence on imports will be 64%, compared to 42% at present. The price of crude in 1990 constant dollars is expected to be at $34 per barrel in 2010, 70% above the 1991 price.

For California and the West Coast, crude supplies are expected to decline due to lower production from Alaska, Elk Hills, conventional (light) crudes in California, and to environmental politics blocking oil leasing, exploration, and production from marine resources. Only heavy oil production from Kern County is expected to increase, and that is expected to peak out in 1995.

With some increase in the West Coast demand for crude, the crude oil surplus characteristic of the West Coast will disappear in 1994, according to our forecast. In that year, imports will rise to fill the deficit, and increase steadily thereafter. The results are that in 1994 crude oil prices on the West Coast will rise by about $2 per barrel (10%), and oil product prices will increase proportionately; and that the West Coast will join the rest of the nation in becoming increasingly dependent on imports for vital crude oil supplies.

PUBLIC POLICY ISSUES

Looking into the next century, we find that declining U.S. production of crude oil and increasing dependence on Middle Eastern oil, followed by declining world oil reserves and rising oil prices, are high-probability events. What public policy issues may be considered to deal with these events?

Should the U.S. Strategic Petroleum Reserve be expanded?

The Strategic Petroleum Reserve (SPR) is a means of providing a high degree of supply security, given that the U.S. will become increasingly dependent on foreign crude oil. The supply security issue is the subject of Chapter 3. Here we will merely note that a satisfactory solution exists at reasonable costs.

The options consist of salt dome, abandoned mine, steel tank, and *in situ* storage in which an existing oil field is fully developed but shut in (Mead, 1973). The oil storage cost consists primarily of the interest charge on the stored oil, plus a charge for developing and maintaining the storage facility. Some (perhaps all) of the interest cost will be offset by increasing prices of crude oil, which raise the value of the stored crude. Full offset for the interest cost requires that the *in situ* value of oil increase at the opportunity cost of capital (the market rate of interest).

The SPR using salt dome storage was intended to be a 1-billion-barrel system. In fact, the SPR has increases from zero in 1982, to 590 million barrels as of September 1990. It was drawn down 22 million barrels between October 1990 and April 1991 to replace oil lost during the period of the Persian Gulf crisis. A 568-million-barrel reserve is sufficient to replace all OPEC imports into the U.S. for 137 days, at 1989 import levels. Since the loss of all OPEC imports is highly unlikely, the number of days of supply security is well in excess of 137 days. If the SPR is increased to its planned 1-billion-barrel capacity, U.S. oil supply security appears to be fully attainable relative to any likely supply disruption.

Should Elk Hills be converted to an SPR?

All of the salt dome SPR reserves are in the Gulf of Mexico (GOM) region. Supply security for the West Coast would come from diverting to the West Coast some imports normally received in the GOM. As an alternative, the Elk Hills field near Bakersfield, California (85% owned but 100% controlled by the federal government) could be shut in and converted into an SPR (Mead and Sorensen, 1971).

The advantages of using an existing oil field are, (1) that oil need not be lifted from one field, transported, and then reinjected into a storage facility that must be constructed; (2) that the field is fully developed; and (3) that a pipeline system for delivery to markets is in place. The disadvantage is that oil recovery from the field in an emergency would be limited by reservoir characteristics and at most only about 10% of the oil in place could be recovered in any year. In contrast, oil can be recovered from a salt dome, abandoned mine, or steel tank storage within any likely crisis period. The Elk Hills alternative was found to be the least expensive, with abandoned mine storage second, and salt dome or steel tank storage most expensive (Mead, 1973, p. 38).

All during the period of the West Coast oil glut, the federal government elected to produce at full bore from the Elk Hills field, thereby contributing to the glut, while salt dome storage was developed as the primary SPR system. The Elk Hills field has now been drawn down from 1,363 million barrels to 524 million barrels. Consequently, its cost as an SPR would be higher on a per-barrel basis.

Should subsidies for oil producers and for substitute technologies be reintroduced?

Beginning in the 1920, tax subsidies were given to owners of oil reserves (royalty owners) and to producers of oil. These subsidies consisted of the percentage depletion allowance and the right-to-expense intangible drilling costs. Percentage depletion was removed for all integrated oil companies in 1971, and reduced in value for small producers. Rights-to-expense intangible drilling costs still are allowed.

In addition, mandatory import quotas were added in 1959 that had the effect of limiting imports and thereby increasing the value of domestic oil reserves. The quota system was removed in 1971 when it became unworkable.

These subsidies collectively stimulated early production of a nonrenewable resource thus achieving a degree of supply security in the early years at the expense of greater dependence now. Similar policies are advocated today, often rationalized as a means of gaining some supply security. Such policies ignore the facts that the oil resource is nonrenewable and a barrel of oil can be produced only once. Therefore, such policies cannot achieve a long-term supply security goal. While supply security is the common rationalization, the economic effects of the subsidies noted above are to either reduce tax costs or to reduce competition and thereby raise the price of crude. Both results raise the *in situ* value of oil assets within the subsidized or protected political jurisdiction. Thus, they enrich owners of oil resources at the expense of economic efficiency and the public interest.

Similarly, subsidies were introduced during the Carter administration that were designed to bring forth new technologies that would reduce the nation's dependence on imported crude oil. Proceeds from the "windfall profits tax" were earmarked for this and other purposes. Funds were allocated for coal gasification and liquefaction, solar and wind electric power, along with other substitute technologies. Expensive gasification and liquefaction plants were proven to be uneconomic and are now either bankrupt or abandoned. These investments may have yielded some technological spill-over benefits to society but their social costs appear to exceed their social benefits. If public subsidies are foregone, new technologies will automatically draw investments when they appear to investors to be economically viable. At best, public subsidies will bring them on-stream a bit earlier.

Should leasing of California and federal marine lands for oil and gas production be resumed?

Leasing of California marine lands for possible oil (and gas) production was halted after the Santa Barbara oil spill in 1969. As of this moment, no state leasing is underway or planned. The California State Lands Commission in 1989 decided it would not lease any state marine lands in the future and instead would turn all unleased areas into a coastal sanctuary (Los Angeles Times, 1991b).

Leasing of federal lands was resumed after the spill, but in the last decade leasing has come under fire from environmental organizations with the result that leasing has been curtailed and then virtually halted. In June 1990, President Bush announced that in two large West Coast areas (Sale #91 off the Northern California Coast, and Sale #95 off the Southern California coast) drilling would be banned until the year 2000, and drilling in Monterey Bay would be permanently banned. In effect, about 99% of the California OCS will be off-limits to new leasing and drilling.

Whether new leasing is justified depends on the social benefits and costs of developing new information about hydrocarbon reserves in the marine provinces. Oil companies are the best judges relative to the private costs and benefits. The principal external costs are related to environmental costs that have not been internalized. The relevant environmental issues and costs are discussed in Chapter 7. The principal external benefits are related to oil supply security discussed in Chapter 3.

Should the ban against exporting Alaskan oil be eliminated?

The Export Administration Act of 1979 banned the export of ANS crude oil. The Act states that "no domestically produced crude oil transported through the Alaskan pipeline may be exported from the United States." The ban may be lifted only if the President certifies that the export of Alaskan oil is in the national interest and meets several other specified conditions.

This restrictive legislation is the result of a political compromise to satisfy the demands of maritime unions and maritime firms that the government prohibit oil exports that would use lower cost foreign shipping,[2] and the demands of environmental organizations that opposed the construction of the Alyeska pipeline and the consequent required marine transportation of the oil. The maritime unions and firms cite national security benefits due to maintaining a fleet of oil tankers.

The costs of the restriction fall on the State of Alaska, which receives lower royalties and severance taxes based on wellhead values, on the oil company owners of the ANS reserves, on the federal government that receives lower corporate income taxes on the oil profits, and on the public in general that must pay for any and all economic inefficiency.

The primary inefficiency resulting from the export ban is due to distorted shipping patterns mandated by the legislation. Given the fact that the West Coast cannot absorb all of the Alaskan crude, it must be shipped beyond West Coast markets to Panama, off-loaded to be transported via pipeline across Pan-

---

[2]   The Jones Act, enacted in 1926, requires that all marine shipments between any two U.S. ports must be on ships built in the U.S., owned by U.S. firms, and manned by U.S. crews. Without the export ban, shipment of ANS oil beyond California to the Gulf of Mexico (GOM) would be uneconomic. The ANS oil shipped to the GOM would clearly have been exported to the Orient on less expensive foreign ships and replaced with less expensive imports.

ama, reloaded onto tankers, and then transported to ports in the Gulf of Mexico, the Caribbean, or the East Coast.

A recent study by the General Accounting Office (GAO) estimated that lifting the ban would raise the wellhead value of ANS oil by $2.16 per barrel (U.S. General Accounting Office, 1990, p. 9). The ban results in a loss of economic rent, which is a measure of the social cost to the nation. The welfare loss to the nation may be estimated roughly by multiplying the volume of ANS oil shipped out of Alaska annually through 1991 by the welfare loss per barrel, and expressing the loss in terms of 1989 dollars.

Using a 4% real interest rate, the social welfare loss due to the ban on ANS is estimated at $21.6 billion (Alaska Department of Natural Resources, 1991, p. 11). This calculation underestimates the true loss because it is based on the record through 1990, ignoring the projected production for as long as the export ban is in effect. Whether the export ban and its cost to the nation is justified depends on whether the benefits to national security resulting from subsidizing the oil tanker fleet and its union employees is worth more or less than $21.6 billion. We have no estimate of this possible value.

REFERENCES

Alaska Department of Natural Resources, Division of Oil and Gas (1991). "Historical and Projected Oil and Gas Consumption." June.

California Energy Commission (1989). Fuels Report, Appendix, p300-89-018A. December. Sacramento.

Chevron U.S.A., Inc. (1991). "Crude Oil Price Bulletin." 91-CA12. San Francisco.

Conservation Committee of California Oil and Gas Producers (1990). "Annual Review of California Oil and Gas Production." 1990 and earlier years. Los Angeles.

Electric Power Research Institute (1991). "They're New! They're Clean! They're Electric!" *EPRI Journal*, Vol. 16, No. 3, pp. 4–15. Palo Alto, Calif.

Guerard, W. F., Jr. (1984). "Heavy Oil in California." Publication TR 28, Third Edition. Sacramento: California Department of Conservation, Division of Oil and Gas.

Hogan, D. A. (Director, Naval Petroleum Reserves in California, U.S. Department of Energy) (1990). Private communication, December 12.

Ibegbulam, B. (1991). "California Demand and Supply of Crude Oil: An Econometric Analysis with Projections to 2000." Ph.D. dissertation. June. Santa Barbara: University of California.

Los Angeles Times (1991a). "Unocal to Close the Nation's Last Shale Oil Project." March 27, Sec. D, pp. 1, 4.

Los Angeles Times (1991b). May 12. Sec. A, p. 30.

Masters, C. D., D. H. Root, and E. D. Attanasi (1990). "World Oil and Gas Resources: Future Production Realities." *Annual Review of Energy*, Vol. 15, pp. 23–51.

Mead, W. J. (1973). "Strategic Petroleum Reserves." Hearings before the Committee on Interior and Insular Affairs, U.S. Senate, 93rd. Cong., 1st. Session, May 30, 1973, pp. 19–58. Washington, D.C.: Government Printing Office.

Mead, W. J. (1989). "The Supply of Heavy Oil from California in the 1990s and Beyond." *Energy Supply in the 1990s and Beyond.* Proceedings of the 11th Annual Conf. of the International Association of Energy Economists, Caracas.

Mead, W. J. and M. Denning (1991). "Estimating Costs of Alternative Electric Power Sources for California." In R. J. Gilbert, ed., *Regulatory Choices: A Perspective on Developments in Energy Policy*, pp. 187–259. Berkeley: University of California Press.

Mead, W. J. and P. E. Sorensen (1971). "A National Defense Petroleum Reserve Alternative to Oil Import Quotas." *Land Economics*, Vol. 47, pp. 211–224.

Oil and Gas Journal (1989). September 18. Tulsa, Okla.: Petroleum Publ. Co.

Oil and Gas Journal (1990a). December 31. Tulsa, Okla.: Petroleum Publ. Co.

Oil and Gas Journal (1990b). Various issues, 1990 and earlier. Tulsa, Okla.: Petroleum Publ. Co.

Porter, E. D. (1990). "Non-OPEC Supply and World Petroleum Markets: Past Forecasts, Recent Experience and Future Prospects." Research study No. 054. August. Washington, D.C.: American Petroleum Institute.

Riva, J. P. (1991). "Persian Gulf Oil: Its Critical Importance to World Oil Supplies." March. Washington, D.C.: Congressional Research Service, Library of Congress.

Sande, W. N. (1983). "Oil Shale and Bituminous Sands." In Proceedings of the 12th World Energy Conference, New Delhi, September 18–23, pp. 48–53. Delhi: Maya Enterprises.

U.S. Department of Energy, Energy Information Administration (1990a). *International Energy Outlook*. Washington, D.C.: Government Printing Office.

U.S. Department of Energy, Energy Information Administration (1990b). "U.S. Crude Oil, Natural Gas, and Natural Gas Liquids Reserves, 1989." Annual Report DOE/EIA-0216, October. Washington, D.C.: Government Printing Office.

U.S. Department of Energy, Energy Information Administration (1991a). *Monthly Energy Review*, 1991 and earlier years (various issues). DOE/EIA-0035. Washington, D.C.: Government Printing Office.

U.S. Department of Energy, Energy Information Administration (1991b). *Annual Energy Outlook*. DOE/EIA-0383(91). Washington, D.C.: Government Printing Office.

U.S. General Accounting Office (1990). "Energy Security, Impacts of Lifting Alaskan North Slope Oil Exports Ban." Report GAO/RCED-91-21. November. Washington, D.C.: Government Printing Office.

World Energy Conference (1983). Proceedings of the 12th World Energy Conference, New Delhi, September 18–23. Delhi: Maya Enterprises.

ALTERNATIVE
4 TRANSPORTATION ENERGY

**Daniel Sperling and
Mark A. DeLuchi**

INTRODUCTION

Transportation energy issues are moving to the forefront of the public consciousness in the U.S. and particularly California, and gaining increasing attention from legislators and regulators. The three principal concerns motivating interest in transportation energy are urban air quality, oil dependence, and the threat of global warming. Transportation fuels are a principal contributor to each of these. The transportation sector, mostly motor vehicles, contributes roughly half the urban air pollutants, almost one-third of the carbon dioxide, and consumes over 60% of all petroleum.

One promising strategy for resolving pollution and energy problems is the use of clean-burning alternative fuels. Alternative fuels are an appealing technical fix. They require much less change in personal behavior than mass transit and ridesharing, and minimal changes in the behavior and organization of local governments. They relieve the pressure to coordinate and manage growth on a regional level. Alternative fuels are attractive because they are less disruptive

Daniel Sperling and Mark A. DeLuchi are with the Institute of Transportation Studies, University of California, Davis. Portions of this chapter have been published in a report to the California Policy Seminar and Organization for Economic Cooperation and Development. Funding from California Policy Seminar and University of California Transportation Center are gratefully acknowledged.

politically and are institutionally easier to implement than strategies aimed at reducing the use of single-occupant autos and changing land use. Indeed, because they provide the promise of being environmentally benign, alternative fuels tantalize us with the prospect of never having to restrict motor vehicle use.

It can be argued, using practically any set of conceivable assumptions, that the use of large amounts of alternative fuels are inevitable. They are clearly an important part of any long-term solution to urban air pollution, global warming, and diminishing energy security.

But when, where, and to which fuels will this transition occur? The public debate over this transition question has been unusually muddled and distorted. Why is this?

One explanation for the muddled and distorted debate is that many powerful industry groups have a large vested interest in the success (or failure) of one or more options, including the auto and oil companies, petrochemical and coal companies, agribusiness, and the natural gas and electricity industries. All have a lot to lose, or gain. All have high-powered lobbyists and large public relations and advertising budgets. These vested interest groups, operating in the public arena, often exaggerate and spread half-truths, easy to do given the difficulty of evaluating advantages and disadvantages on a common scale. Exaggerations and half-truths are routine and, unfortunately, accepted behavior in public policy debates.

A second explanation is that numerous fuel options are available; each option has a very different set of advantages and disadvantages. No one option is obviously superior to all others. Comparing the different attributes is like comparing apples and oranges. Because these differences are market externalities or related to performance and ease of use, they are not easily converted to monetary terms and therefore not readily comparable. Some alternative fuels help reduce air pollution, some enhance energy security, and some reduce greenhouse gas emissions. Which goal is most important? Should we choose a particular fuel if it significantly reduces urban smog, but doesn't help the other problems? And at what cost? The selection of a particular energy option is based upon world views and fundamental values, as well as technical judgements.

These evaluation difficulties are aggravated by the fact that each option is at a different state of development. Cost, performance, and emission estimates for today's gasoline fuels and vehicles, which have undergone 100 years of intense development, should not be directly compared to cost, performance, and emission estimates of hydrogen fuels and vehicles, which are at a very early stage of development, or even to estimates of compressed natural gas vehicles, which have been retrofitted after-market for several decades but have received only minimal attention from the major auto-makers.

These evaluation difficulties result in technical studies of alternative fuels that often have widely varying conclusions regarding the relative merits of the fuel options.

The goal of this chapter is to untangle technical judgements from the influence of ideology, values, and economic interests. We place narrow findings in a broader systems context, and peel away the layers of self-interested "technical" findings; we specify core knowledge of the relative merits of the principal petroleum substitutes, acknowledging uncertainty where it exists.

The status of the leading alternative fuel candidates is presented in terms of technology, cost, environmental impacts, energy security, and safety implications. We explain why inconsistencies exist and disclose misconceptions.

Analytical Caveats

Where appropriate and unless otherwise specified, the attributes and drawbacks of alternative fuels and their associated technology are compared with unleaded gasoline used in new automobiles that meet current emission standards. A few problems inhibit this type of analysis.

First, there is little uniformity among gasoline blends sold throughout the U.S. and, because gasoline compositions vary considerably according to producer, locality, and time of year, there is no gasoline standard by which to compare alternative fuels. This problem is becoming more severe in the 1990s as oil refiners, responding to new air quality rules, introduce new formulations of gasolines and oxygenated fuel blends that vary from region to region.

Secondly, we must compare gasoline vehicles—a product that has benefitted from decades of research—to relatively new alternative fuel technologies. But even gasoline vehicles are not a fixed target; for instance, even though emissions of gasoline cars were reduced 90% or so from the mid-1960s to late 1980s, new rules in California call for roughly another 80% reduction by the early 21st century. Still, new technologies generally outpace more mature technologies (if sufficient resources are devoted to the task). Therefore, despite the fact that we can only estimate the capabilities of first generation mass-produced alternative fuel vehicles, technological advances for alternative fuels are likely to be swifter than those of conventional petroleum vehicles.

Finally, the scale and rate of alternative fuel implementation will define many of the important parameters discussed in this chapter. We do not discuss implementation strategies nor market potential here, but assume in the evaluation that demand will be sufficient to stimulate large-scale production of alternative fuels and wide-scale installation of refueling infrastructure.

The Non-Problem

The energy problem is *not* that petroleum supplies will soon be used up. Proven reserves of world oil have been increasing steadily, with new discoveries keeping pace with increasing consumption (U.S. Department of Energy, 1992). If one were willing to rely on Persian Gulf countries for their oil supply, and if the Persian Gulf countries could be relied upon to supply oil at their cost of production, there would be no need to worry about oil for many decades. Even if future oil discoveries begin to lag significantly behind consumption, there are many other energy resources that could be used to manufacture transportation fuels.

Indeed, because of the availability of these other resources, it will be a very long time before future prices of transportation energy exceed 1981 oil prices on a sustained basis. Natural gas can be economically used as compressed or liquefied gas or converted into methanol when oil prices are considerably less than $4 per barrel (1988 dollars), the prevailing price in 1981. At about that 1981 price, coal and biomass could be economically converted into methanol, substitute natural gas, and possibly petroleum-like liquids, and oil shale could be processed into gasoline and diesel fuel (Sperling, 1988; National Research Council, 1990). Since natural gas, coal, and oil shale are all available in larger quantities than petroleum, worldwide as well as in the U.S., that means sufficient energy resources are available at or near 1981 prices for at least another century.

After that time, if necessary and if desired, a permanent transition could be made to renewable resources: hydrogen made from water using photovoltaic solar energy, electricity made from solar and other renewable sources, and to a limited extent, liquid fuels made from biomass. As indicated above, biomass fuels will probably cost about the same as coal-based fuels and be environmentally superior, although their production should probably be limited so as not to exacerbate soil erosion and other problems associated with intensive land use (such as loss of biodiversity, and use of fertilizers and water). The (private) production cost of hydrogen is currently much higher than that of other fuel options, but hydrogen does provide non-market benefits of much lower pollution, with good prospects for much lower hydrogen costs in the early part of the 21st century.

The point is that the world is not in imminent danger of running out of energy, and with a well-functioning market system, energy prices will not increase dramatically in the foreseeable future. But the international petroleum market is not a well-functioning market; not only is it erratic and politicized, distorting energy decisions through inappropriate price signals and uncertainty, but it also does not account for large environmental impacts.

SOCIAL AND NON-MARKET COSTS

Design of a transportation fuel strategy should be predicated upon an understanding of the full range of private market costs as well as non-market social costs: private market costs because they are the criterion that industry and individuals use in deciding whether to invest in and purchase alternative fuels, and social costs because they are the justification for government intervention. In the following paragraphs, the importance of air quality, energy security, and reduced global warming are explored.

Energy Security and Petroleum Dependency

The concept of energy security is an autarchic notion that a country should not become excessively dependent on foreign suppliers. Dependency occurs when the good or resource can be acquired more cheaply outside the home country (and government actions do not restrict foreign purchases), is important to the economy, and cannot be replaced quickly in the event of a shortfall. The benefits of buying less expensive goods elsewhere are increased economic efficiency. The costs are those of being unable to respond quickly if foreign supplies are abruptly curtailed or if prices are abruptly increased.

The U.S. is becoming increasingly dependent on oil imports. The trend is unmistakable: domestic oil production is on a downward trajectory and domestic oil consumption is increasing.

In 1990, U.S. crude oil production averaged 7.35 million barrels per day, the lowest in several decades. The U.S. Department of Energy (DOE) projects in its mid-range scenario that domestic oil production will drop another 1.85 million barrels per day by 2010 (U.S. Department of Energy, 1992). At the same time, domestic oil consumption continues to increase, mostly due to increased diesel fuel and jet fuel use. The U.S. DOE forecasts a 3.2 million barrels per day increase in domestic consumption between 1990 and 2010.

As a result of these production and consumption trends, imports are expanding. In 1990, oil imports accounted for 42% of consumption, close to the peak of 47% recorded in 1977. DOE expects this percentage to increase to 53% to 68% by 2010.

The transportation sector, unlike other energy-consuming sectors, has remained almost completely dependent on petroleum fuels. As a result, transportation has gradually increased its share of the petroleum market. In the U.S., transportation increased its share from 53% of petroleum consumption in 1977 to 64% in 1990 (*Ibid.*). In California, transportation accounts for about three-quarters of oil consumption (California Energy Commission, 1991).

Already, the U.S. transportation sector by itself consumes more petroleum than is produced in the entire country. This level of dependency is unlikely to remain acceptable politically and perhaps economically.

The importance of this import dependency problem is unclear. The severity of the problem depends on one's view of the future: Will OPEC be able to regain market control and escalate oil prices? Will Saudi Arabia succumb to revolution? Will radicalized oil producers decide to use oil as a political weapon? Will another war break out in the Persian Gulf area, and with what repercussions? Will Iraqi and Kuwaiti oil production be resumed at pre-1990 levels. The cost of oil dependency is difficult to measure; it depends not only on determinations of the probability of the foregoing types of events occurring, but also on how the cost of military expenditures in the Middle East and other important supply regions are allocated, the cost of maintaining the U.S. Strategic Petroleum Reserve (now containing over 500 million barrels), the risk of supply disruptions, and losses in national income from contraction of demand for U.S. goods and services. The sum of these costs have been estimated to be as high as $21 to $125 billion per year (DeLuchi, Sperling, and Johnston, 1987).

Import dependency will probably not be the principal motivation for initiating a transition to alternative fuels in the near future, even with disruptions such as the August 1990 Iraqi takeover of Kuwait. Oil-import dependency is expected to grow, however, thereby attracting increasing political attention, and creating at least some pressure for the introduction of non-petroleum fuels.

Dependency on oil imports is not just a problem of security, however. It is also a problem of large indirect economic costs caused by price volatility and increasing world oil prices, resulting in increased revenues for exporters and increased costs to importers. The availability of a credible alternative (and/or reduced petroleum consumption) would dampen oil price volatility and restrain oil price increases. Price volatility is due in part to the uncertain cost and availability of still-undiscovered oil, but more so to the concentration of easily accessible (and therefore low cost) oil in a few lightly populated countries. The finite nature of the resource and, for a few fortunate countries in the Middle East, huge supplies of cheap oil, tempts those countries to manipulate oil prices and supplies.

Price volatility creates uncertainty and distorts investment decisions, resulting in a preference for short-term investments. Erratic and uncertain petroleum prices result in wasted investments such as delays in introducing energy-efficient equipment in the 1960s and early 1970s, billions of dollars of losses on over-enthusiastic investments in synthetic fuel plants in the late 1970s and early 1980s, apparently premature "filling in" of oil wells with high production costs in the late 1980s, and missed opportunities to use enhanced recovery techniques to extract oil from existing oil fields.

The absence of a credible alternative to petroleum transportation fuels also results in oil prices being higher than they would otherwise be. This effect holds for the long term as well as in response to rapid price escalations. Initial efforts at modelling the effect of alternative fuels on world petroleum prices indicate that substituting an alternative fuel for 2 million barrels per day of gasoline fuel would lower the world oil price by about $1 per barrel (Difiglio, 1989). Thus the price suppression benefit to the U.S. in 1995 of those 2 million gasoline-equivalent barrels would be about $9 million per day or $3.3 billion per year.

The effect is even more dramatic for short-term price spikes. If, for instance, petroleum prices were to increase quickly to 1981 levels, which is plausible once excess world capacity is used up in the 1990s or later, then oil importers would be faced with steeper spikes that dropped off more slowly than otherwise. If oil importers wait for the higher prices, they will not be able to react with substituted fuels for many years.

High prices could be maintained for 20 years or more as the U.S. and other oil importers struggle to expedite the transition to non-petroleum fuels and to replace vehicles that consume only gasoline and diesel fuel.

Indirect economic costs are a powerful motivation for introducing alternative fuels, but because the costs cannot be accurately quantified and because they are so diffuse, they probably will not play a principal role in motivating the introduction of new fuels.

Greenhouse Effect

A second problem, global warming, is caused by emissions of carbon dioxide and other trace gases that create a greenhouse effect. It attracts more attention than energy security or indirect economic impacts, in part because the potential costs are much greater—though also more speculative.

The scientific community is in agreement that the globe's temperature will increase and climate patterns will change if emissions of carbon dioxide and other greenhouse gases in the atmosphere continue to increase (Science, 1990). Still uncertain is how fast this effect will occur, and how climatic patterns will change. It is expected that the warming will be disproportionately near the poles, eventually causing melting of ice masses and increases in ocean levels. Gradual but ultimately dramatic changes could occur in local and regional climates. Rainfall would increase in some areas, decrease in others, and atmospheric temperatures would change, increasing in most but not all locations. Although these climatic changes can not be predicted accurately with existing meteorological models, it is clear that there is the potential for major environmental and economic damage.

The principal source of carbon dioxide and other greenhouse gas emissions are carbon-bearing fossil fuels: oil, coal, natural gas, and oil shale. Transportation accounts for 34% of the carbon dioxide gases emitted in California. As scientific evidence becomes more certain, the possibility exists that a strong commitment will be made to reduce the use of carbon fuels. It is unlikely that carbon dioxide emissions could be reduced economically using control technologies on vehicles or refineries. The most effective strategies for reducing greenhouse gas emissions from transportation is reduced use of chlorofluorocarbons (CFCs) in air conditioners and less consumption of petroleum, either through fuel efficiency or the use of non-fossil fuels, including biomass, hydrogen made from water with non-fossil electricity, and electricity made from non-fossil fuels.

Air Pollution

The third imperative for introducing alternative transportation fuels is, in the U.S., politically the most potent: air pollution improvement. The use of petroleum for transportation results in large quantities of pollutant emissions from vehicles, refineries, and fuel stations. What makes the air pollution imperative most salient in the public policy arena is the existence of a set of institutions and rules for improving air quality.

Virtually all metropolitan areas of the country experience high levels of air pollution. Roughly 60 to 100 metropolitan areas (representing 80–130 million people) do not meet the statutory ambient air quality standards of the U.S. Clean Air Act for ozone, including all the metropolitan regions in California. In 1988 the State of California, responding to evidence that the health effects of ozone may be even more severe than had previously been thought, established more stringent ambient ozone standards than the federal government (0.09 versus 0.12 ppm over a 1-hour period, with no excedances allowed, versus three excedances per three years allowed in the federal rules).

As shown in Table 4-1 most of the metropolitan areas in California are so far above the ozone standard, and are growing so fast, that they have little hope of attaining the standards in the foreseeable future. These same areas are also in severe violation of the particulate standard and most of them also violate the carbon monoxide standard. These high pollution levels threaten human health and create the risk of federal and state sanctions.

The external (nonmarket) costs of this air pollution are huge: Estimates for the U.S. range from $11 to $187 billion per year, the large range depending mostly on uncertainty of the number of deaths and illnesses due to pollution and the monetary value assigned to deaths and illnesses (DeLuchi, Sperling, and Johnston, 1987). Portney, Harrison, Krupnick, and Dowlatabadi (1989) es-

TABLE 4-1   Percent of Days Over State Standard, 1987 Summer and Winter
            Seasons.

Region	O_3 1-hr, summer	CO 8-hr, winter	PM10[a] 24-hr
South Coast (LA)	90	42	78
SF Bay Area	22	1	37
Sacramento	35	4	23
San Diego	56	1	19
Fresno	59	3	59
Ventura	54	0	25
Kern	61	0	66

SOURCE: California Air Resources Board (1988).
[a] Particulate matter less than 10 microns in diameter.

timated that implementation of the Los Angeles area (South Coast) air quality plan will generate benefits of $1.5 to $7.4 billion per year in that region.

Motor vehicles are a principal cause of urban air pollution. The California Air Resources Board (CARB) (1990) estimates that cars and trucks contributed 43% of the hydrocarbons (also categorized as reactive organic gases), 57% of the nitrogen oxides, and 82% of the carbon monoxide emitted in the major urban areas of California in 1987. [Motor vehicles emit relatively little particulates from their exhaust, but airborne particulates (PM10) are composed of up to 35% aerosols, which are largely the result of atmospheric chemical reactions of the NO_x and hydrocarbons largely emitted by motor vehicles. CARB estimates that over half the PM10 that is directly emitted from anthropogenic sources is dust kicked up by motor vehicle activity on roadways.]

One of the problems to keep in mind in the later evaluation of fuel alternatives is the uncertain nature of estimated air quality impacts. While it is certain that air quality benefits would occur with the use of natural gas, electricity and methanol, data and modelling results are not in agreement on how large those benefits would be, especially for ozone (Murrell and Piotrowski, 1987; Carter, Atkinson, Long, Parker, and Dodd, 1986; Harris, Russell, and Milford, 1988; DeLuchi, Johnston, and Sperling, 1988; Office of Technology Assessment, 1990).

It is difficult and misleading to specify precisely the differences in emissions and air quality impacts between different fuels, especially for ozone. (1) Emission rates are determined by tradeoffs between emissions, costs, performance, and driveability. If a particular fuel is less polluting, then engines will be designed to emit the maximum allowed and will gain the benefit by other means: reducing the cost of pollution control equipment, increasing engine power, etc. Actual emissions will likely vary considerably across vehicle make and model. (2) Pollutant production is sensitive to the air/fuel ratio of engines. If future engines are designed to run "lean" (high air/fuel ratio) to gain higher fuel efficiency, then NO_x levels would be relatively higher and CO and HC emissions and engine power would be lower than an engine operating at stoichiometric ratios, as are most of today's gasoline engines. (3) A distinction must be made between single-fuel optimized engines and retrofitted or bi-fuel engines.

(4) The fuel must be specified since, for instance, some methanol emission data are based on a fuel consisting of 100% methanol, while others assume 10% or 15% gasoline mixed into the methanol; it becomes even more complicated for multi-fuel methanol/gasoline engines since they will operate on varying blends of methanol and gasoline. (5) The ozone formation process is highly complex, and even the most sophisticated photochemical air quality models have error margins of 30% or more (Tesche, 1984). (6) Only in the Los Angeles areas has sufficient meteorological and spatial pollutant concentration data been collected to operate multi-day photochemical airshed models; results from Los Angeles are not generalizable to other regions.

(7) Emission data for dedicated single-fuel compressed natural gas (CNG), ethanol, propane, and hydrogen engines are much sparser and less accurate than for methanol engines. (8) Recent studies suggest for a variety of reasons related to the realism of emission test procedures and widespread engine tampering, that actual hydrocarbon and carbon monoxide emissions from gasoline vehicles are several times greater than tested emissions. For some fuel comparisons, these errors will not alter the relative ratings of the fuels, but for others, especially electric vehicles, where emissions are more accurately known, the huge underestimates of emissions from gasoline vehicles result in studies biased against the alternatives.

The point we are making is that emission and air quality data for alternative fuels are uncertain and should be viewed with a certain amount of skepticism. Still, crude relationships can be drawn with some reliability, as they are later in the report.

Another factor to keep in mind, to illustrate the notion that it is easier and more effective to introduce the technical fix of alternative fuels, are the meager impacts projected for other urban air pollution control strategies. For instance, a current analysis of the emission impacts of various control strategies in the San Francisco Bay Area produced the following results (Harvey, 1990). Providing free mass transit to riders with income of less than $25,000, doubling transit service outside center cities, managing freeway traffic more intensely through use of metering lights, warning signs, and lane direction changes, imposing $1 daily parking surcharges in cities, increasing bridge tolls by $2, and charging a 2 cents per mile surcharge on vehicles, would each reduce hydrocarbon emissions by only 1% to 2.5%. Each of these strategies requires huge subsidies and/or would face major opposition, and yet provide minimal benefits. (It should noted that the emission and vehicle usage impacts of these transportation control strategies is small because of current dispersed land-use patterns. If land-use patterns were reorganized on a regional level to assure coodination in a transportation sense between housing, work, and services, then vehicle drivers would be much more responsive to incentives to share rides and shift to transit.)

As will be shown later, the use of alternative fuels provides the promise of much larger emission reductions. For instance, if all light-duty vehicles were switched from gasoline to electricity, hydrocarbon emissions would be reduced by about one-third (electric vehicle use results in about 95% less hydrocarbon emission and gasoline-powered autos and light trucks emit about a third of total hydrocarbons). The use of methanol and compressed natural gas would provide substantially less hydrocarbon emission reduction, but the point is that alternative fuel use allows for large emission reductions with relatively little change in user behavior.

The problem associated with continued reliance on petroleum fuels, therefore, is not necessarily long-run supply, but rather ignored social costs (especially air pollution and global warming) and economic losses resulting from unpredictable oil prices, inflexible responses to oil price changes, and absence of substitute fuels. Because the price of petroleum does not take into account these social costs and economic losses, and because of the disjointed and conservative nature of transportation energy systems, alternative fuels and increased vehicular efficiency are uneconomically delayed.

In summary, if market mechanisms were operating efficiently, then optimal consumption and production of oil would follow. But that is not the case. Efficiency improvements and alternative fuels are delayed beyond the time when they would otherwise be economically attractive by uncertain and low gasoline and diesel fuel prices that do not reflect their true cost to society.

Moreover, as indicated later, there are also large start-up barriers to alternative fuels. Because of the start-up barriers and a flawed market, new fuels will only be introduced if they receive strong support from government. Significant government intervention will be premised upon the public-good concerns listed above: the greenhouse effect, dependency on foreign oil supplies, economic benefits of lower energy prices, and urban air pollution.

Recent History of Alternative Transportation Fuels

Several transportation energy alternatives have emerged only to pushed from the public spotlight in recent years. In the 1970s and early 1980s, petroleum-like fuels from oil shale and coal dominated public policy and private investment; in the mid to latter part of the 1980s, methanol was the favorite, and in the early 1990s, electric vehicles moved to the forefront.

In the mid-1970s, just after the 1973 Arab oil embargo, nations began searching for ways to attain energy independence. The major non-petroleum domestic energy resources in the U.S. were coal, oil shale, and biomass. Natural gas was virtually ignored since it was considered even more scarce than petroleum. Curtailments of natural gas deliveries to customers in accordance with the U.S. government's allocation scheme during the winter of 1976–77 served to reinforce the notion that natural gas was a scarce resource that should be reserved for winter heating needs (U.S. Department of Energy, 1987, p. 123).

For the transportation sector, the most attractive options seemed to be petroleum-like fuels produced from coal and oil shale, methanol produced from coal, and ethanol made from corn and other biomass. Ethanol was quickly discarded as a major option by most energy analysts for being far too expensive (although not by the agricultural community, who saw ethanol as an answer to excess production and low prices of farm goods).

Methanol was rated below oil shale and other coal liquid options because it would require major changes in motor vehicles and pipeline and fuel distribution systems and would not support existing investments in oil refineries (Kant,

Cohen, Cunningham, Farmer, and Herbst, 1974). At a Fall 1973 conference on Project Independence sponsored by the U.S. Department of Interior, "oil and automotive industry representatives voiced sharp opposition to a national energy program emphasizing methanol rather than synthetic gasoline fuels" (Bechtold, 1987, p. 3). A 1976 report by Stanford Research Institute (SRI) International prepared for the predecessor agency of DOE rated synthetic gasoline a far more promising alternative than methanol, arguing that oil companies would be extremely unlikely to adopt methanol because "production of synthetic crude allows it simply to be added to the natural crudes still available to refineries . . . serving both the needs of oil companies wishing to maintain the usefulness of present investments and insulating the consumer from change" (SRI International, 1976, p. xii).

Virtually all the major energy studies in the 1970s and early 1980s, as well as government energy policy, favored petroleum-like fuels from coal and oil shale (Kant, Cohen, Cunningham, Farmer, and Herbst, 1974; SRI International, 1976; Purdue, 1981). Public and private research and development was heavily weighted toward direct liquefaction of coal (Perry and Landsberg, 1981, p. 248).

Indeed, as late as 1981, only five of the 31 most advanced synthetic fuels projects in the U.S. intended to produce methanol as a primary product, and of those, several intended to co-produce high-Btu pipeline-quality substitute natural gas (Pace, 1981). Two additional projects intended to manufacture methanol but planned to convert the methanol into synthetic gasoline in order to make the fuel compatible with the existing motor vehicle and fuel distribution systems (essentially downgrading the methanol into a lower-octane, higher-polluting fuel, at additional cost). Methanol was a minor consideration well into the 1980s.

In the early 1980s, perceptions began to shift, motivated by two insights. First, the cost of manufacturing petroleum-like fuels was greater than had been anticipated, and second, petroleum-like synthetic fuels did not help reduce persistent urban air pollution. The cost problem became salient as world petroleum prices stabilized and then dropped and as feasibility studies performed by project sponsors for the U.S. Synthetic Fuels Corporation began to indicate that the cost of producing refined shale oil and petroleum-like liquids from coal would be as much as $100 per oil-equivalent barrel in first generation plants (U.S. Synthetic Fuels Corporation, 1985, p. H-10).

Attention began to shift toward methanol because of the relatively advanced state of coal-to-methanol conversion technology, and shortly thereafter because of a growing realization that much more natural gas existed than had been recognized (American Gas Association, 1985). Although estimates of domestic and worldwide natural gas reserves began to be revised sharply upward in 1979, this was not widely acknowledged until several years later. The changed perception of natural gas availability was crucial because methanol can be manufactured more cheaply and cleanly from natural gas than from coal.

Methanol received more attention than other alternative fuels from the mid-1980s until about 1990. The explanation for this attention is the following: Methanol can be made from a large number of materials, many of them available in abundance in the U.S.; it can be made less expensively than most other options; it emits less reactive air pollutants than petroleum fuels; and because it is a liquid and therefore more similar to gasoline and diesel fuel than other leading candidates, it requires less costly changes in motor vehicles and the fuel distribution system.

The air pollution benefits of methanol first gained attention, although as a secondary issue, in the early 1980s. A study prepared for the California Energy Commission (CEC) (Acurex, 1982) played a key role, not because it gained wide circulation, but because it laid the basis for the Commission's organizational commitment to methanol fuel. This landmark study concluded that, given the state's severe air pollution problems, the most attractive use of coal for California was to convert it to methanol for the transportation and electric utility sectors. This study was important because the CEC proved to be the most influential advocate of methanol through the 1980s, their major justification for this advocacy being the air quality argument (Smith, Fong, Kondoleon, and Sullivan, 1984; Three-Agency Methanol Task Force, 1986).

Interest in methanol began to surge around 1985 as methanol proponents shifted their arguments away from energy security, a diminishing concern, to urban air quality, a stubborn problem for which most of the "easy" solutions had already been exhausted. Proponents, especially in California, argued that "the transition to neat methanol fuels for all motor vehicles represents the most significant opportunity for air quality progress which exists between now and the end of the 20th century" (Berg, 1984).

That argument was overstated. It reflected a perception that gaseous fuels and electric vehicles were too different from liquid fuels, requiring too many costly changes in motor vehicles and the fuel distribution system and in consumer behavior to be a widely used fuel (e.g., see California Energy Commission, 1986a, 1986b, 1987)—exactly the same argument that had been used against methanol 10 years earlier.

In the late 1980s, as analysts began to scrutinize more carefully the relative costs, and air quality, energy security, and greenhouse benefits of the alternative fuels, natural gas and especially electricity began to receive more attention. The perception that only a liquid fuel was acceptable slowly eroded. While methanol continues to retain substantial support, the early 1990s has seen the emergence of electric vehicles as an important if not leading option.

The dramatic emergence of electric vehicles was due to a realization that the air quality benefits of methanol were more modest than originally believed, and that electric vehicles provided the potential for much greater air quality improvements. The action that galvanized industry into action was a rule adopted in late 1990 by the California Air Resources Board, requiring that a growing percentage of each automaker's sales in California must be zero-emission vehicles (ZEVs); the percentage was set at 2% for 1998, increasing to 5% in 2001 and 10% in 2003.

A surge of investment in electric vehicles (EVs) was assured when General Motors announced in early 1991 that it had selected an assembly plant in Michigan for production of an electric vehicle (based on their Impact concept car) and expected to begin production in the mid-1990s. This announcement undermined tentative plans by other auto manufacturers, especially those based outside the U.S., to eliminate or water down the ZEV mandate. The commitment of automotive manufacturers to EV production was further enhanced by a growing perception worldwide that emission standards and rules adopted in California will eventually be adopted elsewhere. Manufacturers around the world began a crash EV development program, and in the U.S. an "Advanced Battery Consortium" was formed by the electric utilities, auto manufacturers, and U.S. government (DOE) to accelerate the development of advanced batteries.

COMPARATIVE ANALYSIS OF ENERGY OPTIONS

Considerable space is devoted here to a comparative analysis to demonstrate the distinct advantages and disadvantages of different options. We argue that each of the fuel options analyzed below can be shown to be superior in some situation, but that no one fuel can be identified as superior to all others in all situations. The transportation energy options analyzed here are biomass fuels, methanol made from natural gas and coal, natural gas vehicles, electricity, and hydrogen. These are the most attractive near- and medium-term options. Liquefied petroleum gases (LPG) and petroleum-like fuels made from coal, oil shale, and tar sands are not included in this report.

Petroleum-like Fuels

Petroleum-like fuels derived coal and oil shale are not considered further in this chapter because they have large negative environmental impacts, including higher levels of greenhouse gas emissions, large quantities of solid waste, large water needs, and introduction of additional toxic materials into the ecosystem (Chadwick, Highton, and Lindman, 1987). The fuels would be considerably more expensive than compressed (or liquefied) natural gas and methanol made from natural gas, although proponents claim that their costs can be reduced significantly with intensified research and development efforts, perhaps to as low as $30 per barrel (Lumpkin, 1988; National Research Council, 1990). The final cost would be considerably higher for the U.S., however, because of the large costs required to reformulate the fuel to meet future emission standards and to meet other increasingly stringent environmental restrictions.

Reformulated Gasoline

Reformulated gasoline is also not analyzed here, principally because of insufficient data. Gasoline consists of a large number of different molecular compounds, ranging from very light near-gaseous hydrocarbon molecules to heavy complex molecules. In practice, no two quantities of gasoline are identical; in fact, refiners purposefully create different gasolines for summer and winter, and for certain regions of the country. Reformulated gasoline is gasoline that has been modified to have lower emissions of hydrocarbons, benzene, and other pollutants. Reformulated gasoline was first proposed as an alternative fuel in summer 1989 in response to the growing pressure for cleaner-burning fuels, in particular the July proposal by President Bush to require the sale of alternative fuel vehicles in the nine most polluted cities of the country. In the fall of 1989 in Southern California, ARCO became the first oil supplier to market a gasoline reformulated for lower emissions. They reformulated leaded gasoline, in part by blending in MTBE, an oxygenated derivative of methanol.

The U.S. Clean Air Act amendments of 1991 required reformulation of gasoline in the more polluted cities of the country, and the California Air Resources Board in late 1991 went one step further, with the encouragement of ARCO (and against the opposition of all other major oil companies), in requiring a much more stringent reformulation for gasoline sold in California beginning in 1996. It is claimed that reformulated gasoline will reduce ozone-causing emissions by 30% or so (and other pollutants by varying amounts) and to cost an extra 15 cents per gallon (U.S. General Accounting Office, 1990; Boekhaus, 1990).

Liquefied Petroleum Gases

LPG is the light part of crude oil and the heavy part of natural gas; it represents a small proportion of oil and gas reserves. It is attractive now because of its low price, but if demand increased in the transportation or other fuels markets, this price advantage would disappear. LPG is not considered as anything more than a niche fuel, even by the LPG industry itself.

BIOMASS FUELS

Biological matter (biomass) can be a feedstock for the production of a range of liquid and gaseous fuels. Although biomass has been used to manufacture transportation fuels since the 19th century, major biomass transportation fuel activities were not initiated until the late 1970s, when Brazil and the United States fermented sugar cane and corn, respectively, into ethanol. About 184,000 barrels per day of ethanol were produced as a transportation fuel in Brazil in 1987 (Trindade and de Carvalho, 1989) and about 50,000 barrels per day in the United States. More than 90% of all Brazilian cars were designed to operate strictly on ethanol from 1983 to 1989. In the United States the ethanol is mixed in a 10/90 blend with gasoline so that it can be burned in conventional unmodified gasoline-powered vehicles. Various developing countries have experimented with biomass ethanol, but with much less success.

Biomass fuels are attractive because the feedstocks are renewable and domestically available, and therefore could permanently displace imported petroleum. The use of biofuels in transportation could result in no net CO_2 produced (because the CO_2 is in effect being recycled), provided that the energy used in the manufacture of the biofuels—by farm machinery and fuel conversion facilities, in the making of fertilizers, and so on—is also biomass fuel or non-CO_2 producing. On the other hand, the potential supply of biomass is limited, production of biofuels is costly, and environmental impacts can be considerable.

Feedstocks and Fuel Production

While virtually all current biomass transportation fuel activities involve the fermentation of crops and food wastes containing large amounts of starch and sugar, the more promising option is the use of lignocellulosic material, especially wood pulp. Lignocellulosic material is more abundant and generally less expensive than starch and sugar crops. The most promising processes for converting lignocellulose (hereafter referred to simply as cellulose) into high quality transportation fuels are thermochemical conversion into methanol or hydrolytic conversion into ethanol. Biomass may also be thermochemically gasified and then cleaned and upgraded into a clean high-Btu gas. The production cost and environmental impacts are similar to those of methanol production, and the end-use attributes are identical to those of compressed natural gas (CNG). For simplicity, this latter option is not explicitly treated here.

Unlike other alternative energy options, biomass could not or, more accurately, should not be depended upon as the sole transportation energy source, except perhaps in land-rich Brazil. In the United States, for instance, even if all the wood pulp now harvested by the paper and wood products industries, including logging and mill residues, and all the harvested corn and wheat, were used to make biomass fuels, there would not be enough to satisfy current United States transportation fuel demand. A biomass fuels industry using dedicated biomass energy plantations could increase current yields of wood pulp on forest land tenfold or more, but total production would still be dwarfed by transportation energy demand unless a large proportion of forest land were diverted to biomass energy plantations (or vehicular energy efficiencies were greatly improved).

Sperling (1988) estimated the upper bound of biomass fuel potential in the United States, assuming no major disruption of existing agricultural and silvicultural markets and land management activities, to be about 1.8 million oil-equivalent barrels per day of fuel. Most of this biomass energy was estimated to come from wood plantations; the remainder would come from wood and crop residues, grass crops, peat, and municipal solid waste. The U.S. Department of Energy around 1990 returned its attention to biomass fuel and successfully inserted a strong statement for biomass fuels in the President's 1991 National Energy Strategy. Using optimistic assumptions, the Solar Energy Research Institute of the U.S. DOE and others (Lynd, Cushman, Nichols, and Wyman, 1991) are now estimating ethanol fuel potential in the U.S. (from cellulosic sources) to be several times higher than those of Sperling (1988).

Production Costs

Biomass-derived alcohols now are much more expensive than gasoline on an energy-equivalent basis, and are expected to remain so for the foreseeable future. Ethanol fuel in Brazil is about as costly to manufacture, on an energy basis, as gasoline produced from oil priced at $30 to $35 per barrel (Sperling, 1987; Geller, 1985); in the United States the cost of ethanol made from corn or other fermentable materials is substantially higher (U.S. Department of Energy, 1988).

The cost of converting cellulose to ethanol or methanol cannot be specified as precisely since the technology has not been commercialized, but a reasonable estimate would be a cost similar to that of converting coal to methanol, ultimately $0.65 to $1 per gallon (Lynd, Cushman, Nichols, and Wyman, 1991; see Table 4-5). This plant-gate production cost is equivalent to a retail gasoline price approaching $2 per gallon, since methanol contains only one-half the energy per unit volume as gasoline (two-thirds for ethanol), and the distribution and retailing cost per gasoline-equivalent gallon of ethanol and methanol are about twice that of gasoline. Recent evidence indicates that improvements in cellulose conversion technology may lower production costs (Wright, 1988), but even so, biomass transportation fuels will not be competitive in price with gasoline until oil prices are at least $30 to $40 per barrel.

Ethanol fuel activities are thriving in the United States and Brazil, despite high production costs, because of the political and economic strength of the agricultural and food processing industries. Blends containing 10% ethanol and 90% gasoline accounted for about 7% of all gasoline sales in the United States in 1988. Ethanol exists in the U.S only because of generous federal subsidies of $0.60 per ethanol gallon (equivalent to $0.90 per gallon of gasoline on an energy basis) and additional subsidies from many state governments. These huge subsidies benefit primarily ethanol manufacturers, but also gasohol blenders and corn farmers.

Environmental Impacts of Biomass Fuel Production

The introduction of biomass fuels has the potential to sharply reduce greenhouse gas contributions by the transportation sector and to provide small improvements in air quality. On the negative side, increased biomass fuel production may increase soil erosion, reduce biodiversity, and require large amounts of water, fertilizer and herbicides.

The combustion of biomass fuels would generate large amounts of carbon dioxide, but these emissions would be roughly offset by the carbon dioxide taken out of the air by the biomass plants via photosynthesis. As long as fossil fuels are not used for process heat in the feedstock processing plant and in other steps of production and distribution, biomass fuels would be a highly attractive strategy for reducing global warming. In practice, though, as is currently the situation with ethanol made in the U.S., non-biomass fuels are used throughout the chain of activities. In fact, most ethanol production plants in the U.S. currently burn coal for process heat.

The most troublesome environmental impact of biomass production will be soil erosion. Although there is considerable controversy over the extent of soil erosion, a conservative estimate is that half or more of U.S. cropland is suffering a net loss of soil. The Soil Conservation Service estimates that average erosion on United States cropland due only to rainfall is 4.77 tons per acre per year (Soil Conservation Service, 1978), while others estimate total annual erosion, including wind erosion, to be as high as 9 tons (Larson, 1979). Since only about 1.5 (Pimentel, 1981) to 5 tons of soil form per acre-year (Office of Technology Assessment, 1980, p. 71), soil formation cannot keep pace with these losses.

New land brought into cultivation to produce biomass fuels will be at least as prone to erosion as existing land (*Ibid.*). If marginal lands are brought into cultivation without careful soil management, comparatively large amounts of soil will be lost. In general, proper soil management can greatly reduce erosion, but in practice it is rare, because of ignorance, reluctance to change, and unwillingness to invest in techniques with long-term payoffs. Consequently, extensive cultivation of biofuels threatens to be economically and ecologically damaging.

METHANOL FROM NATURAL GAS AND COAL

As indicated above, methanol was the most widely promoted alternative transportation fuel in the United States during the late 1980s (Gray and Alson, 1985, 1989; U.S. Department of Energy, 1988; Calfornia Energy Commission, 1986a, 1986b, 1987; McNutt and Ecklund, 1986). In this section, the salient aspects of methanol fuel are analyzed.

Feedstocks

At present, economic and environmental considerations favor natural gas (NG) over coal and biomass as a methanol feedstock. The production of methanol from natural gas is much less expensive and produces much less pollution than coal-methanol processes; emissions from NG-to-methanol plants are similar to those of petroleum refineries, while emissions from coal-to-methanol plants are much greater (Sperling, 1988, pp. 316–17). The least expensive natural gas is so-called "remote natural gas" (RNG), gas in foreign (usually Third World) countries remote from readily accessible markets and available at about $1 per million Btu or less. Initially, methanol would be made in these low-cost, gas-rich countries, including many OPEC countries, and imported to the United States.

Methanol imports would do little to enhance U.S. energy security, and in fact could weaken it, because foreign methanol suppliers might be no more secure than petroleum exporters, and because a drop in the price of oil, due to the substitution of methanol for some gasoline, would in some cases shut down high-cost domestic petroleum production. Methanol use would probably also increase U.S. payments to exporters for energy, which would add to the trade deficit (DeLuchi, Johnston, and Sperling, 1988). However, as demand for methanol and for other uses of RNG grows, remote gas will become more valuable, and its price will rise. Eventually, the price will be high enough to make domestic gas, and then coal and biomass, competitive as feedstocks.

Methanol made from natural gas could supplant petroleum fuels for several decades; the precise duration of a natural gas-to-methanol era would depend on natural gas use in other sectors, the number of vehicles switched to methanol, and the success of natural gas exploration and development efforts.

Environmental Impacts

Methanol from natural gas is not a permanently sustainable transportation option, nor is it dramatically cleaner than gasoline. It may, however, be enough cleaner to help some cities in air quality nonattainment areas make progress toward meeting national air quality standards. Methanol also will be much cleaner than diesel fuel, and may be an attractive strategy in some cases for meeting the stringent 1993–94 emission standards for heavy-duty engines (Santini and Schiavone, 1988).

Unburned methanol emissions from methanol vehicles are generally less reactive than the hydrocarbon (HC) emissions from gasoline vehicles, and thus tend to produce less ozone. This promise of reduced ozone is the primary attraction of methanol vehicles; the only other important environmental benefit is likely to be lower emissions of toxic pollutants. Methanol may produce less CO or NO_x (but not both) than gasoline vehicles (Table 4-2); the result will depend on the air/fuel ratio, the type of catalyst materials used in control devices, and state of cold-start technology. Methanol production from natural gas is probably slightly cleaner than petroleum refining. Methanol from natural gas would not reduce emissions of greenhouse gases from the transportation sector, compared to gasoline and diesel-fuel use. Methanol from coal would cause a large increase in greenhouse gas emissions (Table 4-3).

The magnitude of ozone reduction possible with methanol substitution is uncertain; many studies have been conducted, but the results are controversial and difficult to generalize. In the mid-1980s, several researchers concluded that the use of methanol in all highway vehicles would reduce peak 1-day ozone concentrations in urban areas by 10% to 30% (Systems Application, Inc., 1984; Jet Propulsion Lab, 1983; Nichols and Norbeck, 1985). In Los Angeles (and elsewhere), however, the worst smog episodes occur as pollution builds up over several days; in 1986, smog chamber experiments indicated that methanol use may not be as beneficial in multi-day ozone episodes (Carter, Atkinson, Long, Parker, and Dodd, 1986). Subsequent modeling studies at Carnegie-Mellon University found that in the Los Angeles area, the use of 85% methanol/ 15% gasoline (the most likely combination) in all mobile sources (vehicles)

TABLE 4-2   Percentage Change in Emissions from Alternative Fuel Vehicles, Relative to Gasoline.[a]

Fuel	NMHC[b]	CO	NO_x	O_3	SO_x	PM[c]
Methanol (w/catalyst)	−50	0	0	−50	lower[d]	
CNG, LNG (w/catalyst)	−70	−50	0	−70	lower[d]	
Hydrogen (no catalyst)	−95	−99	?	−95	lower[d]	
Electricity/yr 2010 mix[e]	−99	−98	−75	?	−50	+30

SOURCES: From data in DeLuchi, Johnston, and Sperling (1988), DeLuchi (1989), and Wang, DeLuchi, and Sperling (1990).

[a] These are rough estimates only, assuming advanced-technology, single-fuel cars, and emission control and engine operation designed to meet a NO_x standard, which will be the most difficult standard to meet. Does not include evaporative emissions from vehicle or vehicle refuelling, or emissions from petroleum refining and fuel manufacture. Ethanol fuel is not included because of minimal experience and testing with controlled vehicles. In general, ethanol-powered vehicles will have similar emissions to methanol. One difference is in aldehyde emissions, which may lead to the increased formation of another oxidant, peroxyacetyl nitrate (PAN) with ethanol (Tanner, Miguel, de Andrade, Gaffney, and Streit, 1988).

[b] NMHC = nonmethane hydrocarbons (total hydrocarbon emissions less methane, which is nonreactive and hence does not contribute to ozone formation).

[c] PM = particulate matter.

[d] SO_x emissions depend on the amount of sulfur in the fuel.

[e] See Wang, DeLuchi, and Sperling (1990) for details. Based on forecasts and estimates for California of energy mix for electricity generation, emission control technologies deployed on power plants, vehicle emission rates, and electricity consumption rates of electric vehicles.

except motorcycles and planes would result in only a 6% reduction in peak ozone levels (Harris, Russell, and Milford, 1988; Russell, 1987).

If 100% methanol (M100) were used in advanced technology engines with extremely low formaldehyde emissions, ozone would be reduced 9%, compared to an advanced-technology gasoline engine. The 9% reduction with advanced-technology M100 represents 43% of the maximum ozone reduction attainable from motor vehicles; that is, if all vehicle emissions were eliminated, ozone would be reduced 21% (Harris, Russell, and Milford, 1988). A subsequent study questions these findings, arguing that methanol vehicles would emit more NO_x than gasoline vehicles, and more than is assumed by the Carnegie-Mellon researchers, thereby causing ozone levels to increase (Sierra Research, Inc. 1988a, 1988b). In any case, the greatest potential ozone reductions with methanol require the use of M100 and very low formaldehyde emissions. We estimate that the substitution of methanol for gasoline in all motor vehicles

TABLE 4-3   Emissions per Mile of a Composite Measure[a] of Greenhouse
Gases, Relative to Gasoline-Powered Internal Combustion
Engines.[b]

Fuel/Feedstock	Percent Change
Fuel cells, hydrogen with solar power	−90 to −85
Ethanol from wood	−75 to −40
Hydrogen with nuclear	−70 to −10
EVs, natural gas plants	−50 to −25
LPG	−30 to −10
CNG from NG	−20 to 0
Methanol from NG	−10 to +8
EVs, current U.S. power mix	−20 to 0
Gasoline	—
EVs, new coal plant	0 to +10
Methanol from coal	+30 to +70

Source: DeLuchi (1991).
[a] CH_4, N_2O, NO_x and NMHC mass emissions from vehicles are converted into the mass amount of CO_2 emissions with the same temperature effect where "same temperature effect" is defined as the same number of degree-years over a given time period. See De-Luchi (1991) for details.
[b] The analysis considered emissions of CH_4, N_2O, and CO_2 from the production and transportation of the primary resource (coal, natural gas, or crude oil), conversion of the primary resource to transporation energy (e.g., natural gas to methanol, or coal to electricity for battery-powered vehicles), distribution of the fuel to retail outlets, combustion of the fuel in engines, and the manufacture of the vehicles.

would result in a maximum reduction in peak ozone levels of 0% to 15% in multi-day smog episodes.

Two cautionary notes in interpreting these air quality analyses. One, ozone air quality models are subject to considerable uncertainty because of inadequate input data, especially outside Los Angeles, and two, optimized single-fuel engines are much cleaner-burning than multi-fuel engines.

This second point is critical because the preceding assessment of emission impacts of alternative fuels was based on the assumption that the engines were designed specifically for those fuels. Commercial versions of such optimized single-fuel engines do not yet exist. Indeed, there is relatively little experience with optimized alternative fuel engines and catalyst technology. If a serious sustained effort were made to reduce emissions, similar to the 25-year history with gasoline engines, major emission reductions would be likely.

In contrast to the uncertainties surrounding the environmental benefits of substituting methanol for gasoline, there are several clear environmental advantages to using pure methanol in heavy-duty engines. Methanol produces essentially no particulates, smoke, SO_x, or unregulated pollutants. In addition, an M100 methanol engine with an oxidation catalyst produces very little CO, HCs, and formaldehyde (Ullman and Hare, 1982; Alson, Adler, and Baines, 1989).

In summary, methanol use would not reduce greenhouse gas emissions, but would provide some air quality benefits when used in diesel engines; it may lead to a minor reduction in either NO_x or CO emissions in spark-ignition engines (and perhaps an increase in the other), and has the potential in some regions for achieving a part of the maximum ozone reduction attainable through changes in the transport sector. But the magnitude of these potential improvements is modest.

Safety and Toxicity

One of the primary concerns about methanol has been its toxicity and safety. Methanol causes blindness if drunk, burns with an invisible flame (making it difficult to detect fires), and is highly soluble in water (making it difficult to contain a spill).

The first two of these problems are solved by adding 10% to 15% gasoline (or some other combustible denaturant) to the methanol, making the flame visible and giving the liquid a very unpalatable smell and taste—although this reduces methanol's air quality advantages. The third issue, solubility of methanol in water, is not necessarily a disadvantage; the greater solubility causes the methanol to quickly dissolve, thus not causing the long-lasting destruction typical of large oil spills. Overall, gasoline is a more threatening fuel than methanol: it is far more flammable and contains many carcinogens.

Costs

Methanol is more expensive than gasoline on an energy-equivalent basis, and will continue to be so for the foreseeable future. The most recent estimates are that very small amounts of methanol can be delivered to the United States for as little as $0.20 to $0.30 per gallon if the remote natural gas (RNG) feedstock is virtually free and sunk costs in the methanol plant are ignored (U.S. Department of Energy, 1988). A more reasonable estimate, based on sustainable rate-of-return conditions and assuming competition for the RNG feedstock—including both domestic uses and other exporting possibilities—is $0.40 to $0.60 per gallon (equivalent on an energy basis to $0.80 to $1.20 per gasoline gallon) (*Ibid.*). Methanol could be produced from coal in the United States for around $1 per gallon (Sperling, 1988). When transportation, storage, and retailing costs are considered, methanol from RNG would not be competitive with gasoline until gasoline sold for $1.10 to $1.70 per gallon, including taxes (and allowing for the fact that methanol is about 10% to 20% more efficient than gasoline in internal combustion engines). Methanol from coal would not be competitive until gasoline sold for at least $2 per gallon.

From a public policy perspective, a more relevant analysis might be methanol's cost-effectiveness in reducing ozone pollution, relative to other pollution-reduction strategies. Such an analysis conducted by the Office of Technology Assessment (1989) came to a mixed conclusion.

Their analysis assumed the following: an ozone-reduction potential of methanol relative to gasoline ranging from a low of 30% using M85 to as high as 90% for M100; a cost of $0.05 to $0.56 more per gasoline-equivalent gallon for methanol than gasoline; and an additional cost of zero to $1,000 for a methanol car over a gasoline car. They conducted the analysis for a vehicle that travels 26,000 miles per year (more than twice the national average). The result was that the use of M85 would cost $9,000 to $66,000 to eliminate 1 ton of "ozone-equivalent" hydrocarbon emissions; if M100 were used, assuming favorable ozone-reduction parameters, the cost would be $3,000 to $22,000 per ton.

A similar analysis conducted for California under the auspices of a blue-ribbon advisory board estimated the cost-effectiveness of M85 at $8,000 to $40,000 per ton (California Advisory Board on Air Quality and Fuels, 1989); and a study by Resources for the Future (Krupnick and Walls, 1992) for the U.S. estimated the cost-effectiveness for M85 in 2000 at $33,268 and for M100 in 2010 at $59,736.

Most of the non-methanol ozone-reduction strategies studied by OTA had cost-effectiveness reductions of $500 to $6,000 per ton. While methanol may be less cost-effective than other options it, along with other alternative fuels, nonetheless provides the potential for much larger ozone reductions than any other strategy.

The OTA estimates suggest that *multi-fuel* methanol cars are clearly not a cost-effective ozone control strategy. Given the range of uncertainty in costs and emission reductions, a similarly definitive conclusion regarding optimized dedicated methanol cars is premature since these cost-effectiveness analyses are too narrow, too short-sighted, and highly sensitive to several key parameters that cannot be accurately specified. Cost-effectiveness analysis will almost always tell you not to do something new and unique, because it does not capture all the direct benefits, much less the secondary benefits. And it ignores the fact that with new technologies—such as computers, freeways, recycled paper—we gradually shift our investments and institutions and behavior to accommodate and support those new technologies, and thereby gradually improve their apparent cost-competitiveness.

In any case, if methanol fuel and vehicle prices are not too much higher than those for their gasoline counterpart, and continued advances are made in emission controls of methanol vehicles, then dedicated methanol vehicles could be a cost-effective strategy for reducing ozone.

Opportunities for Methanol

An important first use of methanol (and natural gas) fuels may be in heavy-duty diesel engines. New emission standards requiring sharp reductions in particulate and NO_x emissions from heavy-duty diesel vehicles take effect in the United States in 1994 (1993 for transit buses). Meeting the standards by applying control technology to diesel combustion is expensive, though probably no more so on a lifecycle cost basis than using methanol or natural gas. Several heavy-duty engine manufacturers are developing methanol (and natural gas) engines. These engines have potentially lower particulate and NO_x emissions than controlled diesel engines and will probably be used mostly in those areas with more severe air pollution and/or where regulators and legislators mandate their usage.

However, diesel-powered trucks and buses consume only about two of the 15 quadrillion Btus of energy used annually on the highways in the United States (Holcomb, Floyd, and Cagle, 1987) (although the proportion is increasing). If methanol is to replace a significant amount of petroleum transportation fuel, and have a discernible impact on air quality, it must penetrate the market for light-duty (gasoline) vehicle fuels. A strategy to introduce methanol in this market must address the high cost of methanol fuel compared to gasoline and the large initial costs both for manufacturing methanol fuel and methanol vehicles, and for establishing a national methanol distribution network for light-duty vehicles. The large initial costs and uncertain market create a need for cooperation between fuel producers and vehicle manufacturers.

The problem of fuel cost is straightforward. Consumers will not use methanol, nor manufacturers make dedicated methanol vehicles, unless methanol use is mandated or subsidized to bring its cost below that of premium gasoline. Government perhaps could justify subsidies or mandates on air quality grounds, but not, as noted above, on global-warming or energy-security grounds.

The problem of start-up costs is more complicated. Because of large start-up costs, manufacturers will not invest in the manufacture of methanol vehicles if the methanol fuel is not available, and fuel producers will not invest in the production and distribution of methanol, even when it is cheaper than gasoline, unless there are vehicles that can burn it. To use methanol, motor vehicles must be modified; the cost of building these modified vehicles will be large initially, since retooling and research and development costs must be spread over a relatively small number of vehicles (although at full production the cost of a methanol-powered vehicle is expected to be about the same as the cost of a comparable gasoline-powered vehicle).

Similarly, establishing a methanol fuel delivery infrastructure will be fairly expensive. The minimum cost approach for a large scale effort would be to market the fuel only in and near ports with ocean access, obviating the need to modify the existing oil product pipeline network or to build an entirely new pipeline network. (Since methanol will be imported initially, a port-based distribution system will be adequate at first.) DOE estimates that the additional capital cost of building a national methanol distribution system to replace 1 million barrels per day of petroleum fuels, using only waterborne and truck transport, and with methanol marketed only within 100 miles of major river and ocean ports (reaching about 75% of the United States), would be $5 billion (U.S. Department of Energy, 1990).

The "chicken-and-egg" dilemma created by these large start-up costs could be resolved by coordinating vehicle manufacture, fuel distribution, and fuel production. Such coordination probably would be arranged by state or federal government. Incentives, not necessarily financial, would need to be offered to vehicle manufacturers to induce them to manufacture and market methanol vehicles, and financial subsidies would need to be offered to retail fuel stations and consumers, at least initially, to overcome the price disadvantage of methanol. (We note, however, that what government invokes, it can revoke, and that even with incentives and subsidies, the private sector runs some risk.) Relaxation of vehicle fuel-efficiency standards for manufacturers that market methanol vehicles, as provided for in the Alternative Motor Fuels Act of 1988 (PL 100–494), might be sufficient to induce manufacturers to produce methanol (or other non-petroleum) vehicles.

Retail fuel suppliers will require more direct subsidies, such as the $50,000 capital grants offered by the Canadian government to retail fuel stations to install facilities for compressed natural gas (CNG) and the per-gallon subsidies provided by the California Energy Commission to methanol fuel suppliers.

The combination of subsidies to gasoline marketers to sell methanol, incentives to manufacturers to sell fuel-flexible (as well as dedicated) methanol vehicles, and subsidies to vehicle owners to purchase methanol cars, as has been occurring in California in the early 1990s, may prove effective in overcoming the "chicken-and-egg" dilemma. Ultimately, though, methanol fuel itself would have to be subsidized to convince consumers to buy methanol, since methanol will cost more than gasoline until oil prices reach at least $30 per barrel on a sustained basis. Even with subsidies, oil marketers will be reluctant to provide methanol pumps if fuel-flexible vehicle owners purchase only gasoline—and, of course, at some point it will become politically untenable to subsidize oil companies if they are not selling much methanol.

In summary, because methanol offers modest environmental benefits at modest cost, a long-lasting transition to methanol will occur only if reducing energy imports, slowing the greenhouse effect, and significantly improving air quality are not high priorities. Other options provide greater non-market social benefits.

METHANE FUELS

Feedstocks

Natural gas, comprised mostly of methane, need not be made into methanol to be used as a transportation fuel—it can be stored on board a vehicle in compressed (CNG) or liquefied (LNG) form, and burned in the engine as a gas.

Later, as the availability of natural gas diminishes and its cost increases, a substitute ("synthetic") natural gas could be produced from coal (or perhaps biomass). The principal advantage of this methane path is lower fuel cost to the end user during the natural gas era, because, as explained below, it is cheaper to compress or liquefy natural gas than to convert it to methanol. Methane could remain as an important or even dominant fuel after natural gas supplies become scarce by converting coal to substitute natural gas (mostly methane); the cost for converting coal to methane would be about the same as converting it to methanol. The principal disadvantages of the CNG/LNG path are those associated with storing gaseous fuels in vehicles and establishing a network of retail fuel outlets.

In the U.S. remote natural gas (RNG), unlike methanol, will not be a major feedstock for NG transportation fuels, unless the cost advantage of remote natural gas feedstock increases, or there is large demand for LNG by LNG vehicles. This contrasts with the methanol case, in which remote natural gas will be a more economical feedstock than domestic gas. (That it is more economical to make methanol from remote than from domestic gas, but more economical to make CNG from domestic gas than from remote gas, is due to the fact that in the methanol case the cost advantage of the cheaper remote feedstock relative to domestic feedstock must compensate only for higher transportation costs, but in the CNG case must compensate for the cost of liquefaction and regasification as well as for higher transportation costs. There is, in other words, an extra step in the RNG-LNG-CNG route—namely, LNG—compared to RNG-methanol, and this extra step is costly enough to tip the economic balance away from RNG.)

This difference—that methanol will be made initially from foreign gas, whereas CNG or LNG will be made from North American gas—may give CNG and LNG an edge in "energy security." The total amount of fuel imports, and the total risk of disruption and outflow of funds, would be lower with natural gas fuels than with methanol.

Another resource consideration is that domestic natural gas resources will last somewhat longer if used as CNG or LNG than as methanol, because conversion losses are much less. We estimated energy losses during each of the following activities: recovery of natural gas (95% efficient), transmission and distribution of natural gas and finished product (95% efficient), reforming of NG to methanol (68% efficient), and NG liquefaction (80% efficient) or compression (94% efficient) (DeLuchi, Johnston, and Sperling, 1988). Based on these estimates, the overall energy efficiency of the NG-to-CNG chain is about 85%, compared to 61% for NG to methanol, and 72% for NG to LNG.

Natural Gas Vehicle Technology

Internal-combustion engines may be readily adapted to operate on CNG. They may be retrofitted, as are all but about 30 of the 500,000 or so CNG vehicles currently operating worldwide, at a cost of about $1,500 to $2,500 per vehicle. The major change is the addition of one or more pressurized tanks for compressed natural gas (CNG) storage, additional fuel lines for the gaseous fuel, and a gaseous fuel mixer in the engine. A far superior vehicle would be one designed specifically for natural gas and not burdened by redundant fuel systems. A vehicle dedicated to and optimized for natural gas would have generally lower emissions than gasoline vehicles, about 10% greater efficiency because of its higher octane and power, and would cost about $700 to $1,000 more because of the more costly fuel tanks, but would not have cold start problems. It would also have a shorter driving range or reduced trunk space because of the much lower volumetric energy density of gaseous fuels (see Table 4-4).

Methane can be stored in carbon skeletal networks called adsorptents. The potential advantage of adsorption is that a given energy density can be attained at a pressure lower than that required to compress natural gas by itself to the same volumetric energy density. For example, an adsorptent at less than 1000 psi can attain the same volumetric energy density as CNG at over 1500 psi. This form of storage, although not yet commercially viable, may lower the cost and bulk of storing natural gas, and may make low-pressure home compression viable. In the United States the Gas Research Institute is sponsoring research and development work aimed at commercializing adsorptents.

Currently, large numbers of CNG vehicles are operating in Italy, New Zealand, Canada, and the former Soviet Union countries (Sathaye, Atkinson, and Myers, 1989). All are retrofitted gasoline-powered vehicles. About 300,000 vehicles have been operating since the 1950s in Italy, mostly in fleet use. Governments in the remaining three countries initiated major CNG programs in the 1980s. In New Zealand, about 110,000 vehicles were converted to CNG, representing at its peak roughly 10% of gasoline use. When the country shifted much of its economy from the public to private sector in the late

1980s, the government withdrew the substantial subsidies it had offered to consumers, and market penetration dropped below the 10% level. The federal and provincial Canadian governments and local gas utilities offered major incentives to fuel suppliers and consumers beginning in the mid-1980s; by 1988 about 15,000 vehicles were operating on CNG, about half by households and half by fleet operators. The Soviet Union had announced the intention in 1988 of converting 500,000 to one million vehicles to CNG by 1995, most of them taxis and trucks, although this plan has apparently been abandoned or dramatically curtailed.

CNG has an extraordinary safety record in actual experience. In New Zealand, for instance, with over 100,000 vehicles in operation for almost 10 years, there has reportedly been only one explosion or fire of a natural gas tank, and

TABLE 4-4   Characteristics of Vehicular Energy Storage Systems.

Vehicle[a]	Range (miles)	Total wt. (pounds)	Total size (gallons)	Fuel dispensing time (minutes)
Gasoline[b]	300	80	11	2
Methanol[c]	300	135	17	3–4
Ethanol[d]	300	110	13.5	2-3
LNG[e]	300	130	27	2–4
CNG/3000 psi[e]	300	240	45	4–8
Liquid hydrogen[f]	300	100	72	3–4
Fe-Ti hydride[f]	150	640	37	5–20[g]
EV/Na-S[h]	150	7000	77	20[i]–720
EV/fuel cell[j]	250	200	80	2–4

[a] The baseline gasoline vehicle gets 30 mpg, lifetime average. Efficiency of other vehicles is referenced to this gasoline vehicle baseline.
[b] 23 lb. gasoline tank, 6.18 lbs./gal., 1.07 outer tank/inner displacement ratio.
[c] 64,000 Btu/gal (cf. 124,000 for gasoline), 6.6 lbs./gal., 15% thermal efficiency over gasoline, 37-lb. tank, 1.07:1 outer/inner ratio.
[d] 84,600 Btu/gal., others as for methanol.
[e] Adapted from data in DeLuchi, Johnston, and Sperling (1988). Assumes fiberglass-wrapped aluminum CNG cylinders; 15% thermal efficiency advantage for CNG and LNG; weight penalty for CNG. LNG system size includes LNG pump.
[f] Adapted from data in DeLuchi (1989). Assumes 25% thermal efficiency advantage; weight penalty for hydride. LH2 system size includes pump. Fe-Ti = iron-titanium.
[g] 80% of hydride refilled in under 10 minutes.
[h] 35-kwh capacity, 120 wh/l, 110 wh/kg, 4.4 mi/battery-kwh (DeLuchi, Wang, and Sperling, 1989). Na-S = sodium/sulfur couple.
[i] Although fast electric vehicle charging is possible, requiring a very large current, it is possible only with certain batteries; development efforts are very preliminary.
[j] DeLuchi, Larson, and Williams (1991).

no one was hurt. The only danger is the accidental leakage of gas from CNG in an enclosed space (in an open space the gas evaporates quickly causing no problems), but again the safety record of CNG in Italy, New Zealand, and Canada has been virtually unblemished. Liquefied natural gas use would be similarly safe since the gas evaporates quickly, unlike gasoline and LPG, minimizing the possibility of fire. LNG could be a problem in enclosed spaces, where leaking or intentionally boiled-off gas would collect, but boiled-off gas could be burned with a small pilot flame, as with a kitchen stove, and rules could be enforced requiring proper ventilation in enclosed garages.

Costs

CNG made from domestic natural gas will be less expensive than imported methanol made from RNG, and much less expensive than methanol made from domestic NG. Landed methanol will cost between $0.40 and $0.50 per gallon, at relatively low levels of demand for the remote NG feedstock, if the low production-cost estimates prove correct. Transport, storage, and retail station costs will add at least $0.14 per gallon to the price, bringing the retail cost to at least $9 per million Btu before taxes, assuming a landed cost of $0.45 per gallon. At the same time domestic gas will be delivered to stations for about $5 per million Btu, according to price projections for commercial gas (U.S. Department of Energy, 1991).

Based on a review of the literature, and a detailed accounting of all costs, including land, site preparation, hook-up to the gas main, energy needed to compress gas from pipeline pressure to 3000 psi, etc., the cost of compressing and retailing CNG is estimated at about $3 per million Btu (DeLuchi, Johnston, and Sperling, 1988; Cann, 1988). Thus a mid-range estimate of the cost of CNG is $8 per million Btu before taxes; a low-end estimate for methanol is about $9 per million Btu. LNG will cost about the same as CNG.

However, because of the high cost of high-pressure storage tanks for CNG, natural gas vehicles would cost about $1,000 more than gasoline and methanol vehicles with the same range and performance. This higher up-front cost is partially compensated by lower back-end costs—the higher salvage value of the storage systems and the probably longer life due to CNG use—thereby increase the resale value of the vehicle.

Ownership and operating costs can be combined and expressed as a total cost per mile over the life of a vehicle, by amortizing the initial cost at an appropriate interest rate, adjusting for salvage values and vehicle life, and adding periodic costs such as maintenance, fuel, insurance, and registration. Table 4-5 presents the life-cycle cost of various alternative fuel vehicles relative to a comparable, baseline gasoline vehicle. It shows the retail price per gallon of gasoline (including taxes) at which the life-cycle cost of the alternative fuel vehicle and the comparable gasoline vehicle would be equal. This is called the "break-even" price of gasoline.

The analysis in Table 4-5 is conducted from the end-user's perspective. The following assumptions are made: the automobiles are optimized for methanol (M100), CNG, and electricity; the fuels are produced and used on a large scale; refueling station costs are fully incorporated; and costs are calculated on a per-mile basis to take into account differences in total life-cycle vehicle costs, including differences in thermal efficiency, maintenance, and engine life. For specific assumptions and documentation, see DeLuchi, Johnston, and Sperling (1988).

The assumptions are based on a review of the literature, including experiences in Europe, Canada, New Zealand, and the U.S., and extensive discussions with vehicle and equipment manufacturers. The analysis is based on a near-term scenario for single-fuel vehicles optimized to run on their respective fuels. The costs associated with CNG vehicles are somewhat more uncertain than those for methanol since the development of CNG vehicle technology has lagged; relatively little effort has gone into designing and testing a vehicle optimized for CNG, including the development of advanced storage tanks, and there is little reliable evidence from which to estimate the operating costs and life of such an optimized vehicle.

The baseline gasoline vehicle, against which the alternative fuel vehicles are compared, has the following attributes: 35 MPG, 2530 pounds, 262-mile range, and a vehicle life of 130,000 miles at 10,000 miles per year. It is assumed that a methanol car costs the same as a gasoline car and that a CNG car costs $1,000 more. The retail price of gasoline, including taxes, is assumed to be $1.15 per gallon, compared to an estimated $0.74 to $1.13 per gallon for methanol and $8.90 to $14.10 per million Btu for CNG. The cost parameters and vehicle attributes are fully documented in DeLuchi, Sperling, and Johnston (1987) and DeLuchi, Johnston, and Sperling (1988).

The methanol and CNG cars are comparable to the baseline gasoline vehicle; they have the same size, range, and weight (excluding the extra weight

TABLE 4-5   Life-Cycle Break-Even Gasoline Prices.[a] (1985 dollars per gallon)

Vehicle/ Feedstock	Fuel Cost	Extra Cost of Fuel Storage[b] ($)	Lifecycle Cost ($)
CNG/ domestic gas	$4–6/mmBtu at station	1000	0.50–1.90
LNG/ domestic gas	$4–6/mmBtu at station	850	0.40–1.80
Methanol/ remote gas	$0.30–0.65/gallon, Calif.	50	0.95–1.70
Methanol/ coal	$0.70–1.00/gallon	50	1.60–2.30
Electric[c]	$0.05/kWh at outlet	7500	0.50–3.50
Electric[c]	$0.10/kWh at outlet	7500	1.00–4.10
Electric[c]	$0.15/kWh at outlet	7500	1.20–4.80
Hydrogen/ fuel cell[d]	$0.05–0.15/kWh on site[e]	2500	1.40–3.20
Hydride/ solar	$0.05–0.15/kWh on site[e]	2600	3.00–12.40
Liquid H_2/ solar	$0.05–0.15/kWh on site[e]	1450	2.80–13.50
Gasoline/ coal[f]		0	1.40–2.30
Gasoline/ oil shale[f]		0	1.60–2.30

[a] The important baseline assumptions used here are: 9% real interest rate for auto loans, 30 MPG, $11,500, 120,000-mile life baseline gasoline vehicle; range and fuel system assumptions consistent with Table 4-4; a range of fuel storage costs are used for each option; methanol vehicles are assumed to have same maintenance costs and life as gasoline vehicles; electric vehicles are assumed to have 25–100% longer life and 25–50% lower maintenance costs; natural gas vehicles are assumed to have 0–20% longer life and 0–15% lower maintenance costs; hydrogen vehicles are assumed to have plus or minus 10% of the maintenance costs and minus 5% to plus 20% of the life; all vehicles are assumed to be optimized for one fuel and produced in high volume. The break-even price of gasoline for a particular alternative is the retail gasoline price that equates the full life-cycle cost per mile of the alternative with the full cost per mile of a comparable baseline gasoline car. Taxes are included. See DeLuchi, Johnston, and Sperling (1988); DeLuchi, Wang, and Sperling (1989); and DeLuchi (1989) for details.

TABLE 4-5   (continued)

b This is the cost of high-pressure gaseous-fuel tanks, cryogenic tanks, liquid alcohol tanks, batteries, or hydrides (for systems with attributes as in Table 4-4), less the cost of the gasoline tank in the baseline gasoline vehicle. A wide range of cost estimates were used for the lifecycle cost calculations. The battery costs have been updated (increased) from the original published estimates.
c Any feedstock from which electricty can be produced and distributed for between 5 and 15 cents/kWh.
d Based on DeLuchi and Ogden (1992).
e Estimated cost of photovoltaic electricity at production site.
f Based on Sperling (1988) and National Research Council (1990).

for CNG tanks and methanol fuel), and similar power. They are assumed to be 10% to 20% and 10% to 25%, respectively, more fuel efficient than the baseline gasoline car.

The lifecycle cost of a CNG auto tends to be less than for a methanol vehicle, although not for all assumed values. The ranges in values correspond to uncertainties in cost estimates and vehicle attributes.

Similarly, CNG vehicles will probably prove to be a more cost-effective strategy for reducing ozone than methanol. The Office of Technology Assessment (1989) report cited earlier estimates that the cost-effectiveness of reducing ozone using dedicated NGVs would be $0 to $1,400 per ton—significantly lower than the $3,200 to $22,000 estimated for comparable single-fuel methanol cars.

Opportunities for Natural Gas Vehicles

Natural gas vehicles have lower fuel costs (per mile), but higher initial cost, than methanol and gasoline, and their refueling facilities are much more expensive than methanol and gasoline fueling facilities. For these reasons, the initial market for natural gas vehicles consists of large fleets with central fueling facilities that use their vehicles intensively. Intensive use is attractive because it takes advantage of the low fuel cost; large centrally fueled fleets are attractive because of the large investment in refueling facilities and the initial absence of a network of public refueling stations.

Beyond this niche, the success of natural gas vehicles depends upon automotive manufacturers participating in their production. Through the early 1990s, all but a handful of the 500,000 CNG vehicles in the world were gasoline vehicles that had been retrofitted to natural gas use. The conversion is expensive, and the resulting fuel efficiency, performance, and emission rates are inferior to those of a vehicle designed specifically for natural gas. Initial efforts by the natural gas industry in the early 1990s to put together an order of several thousand CNG vehicles for General Motors could be the important first step in eliciting manufacturer participation.

The "chicken-and-egg" start-up problem is more difficult for CNG than methanol for two reasons. First, fuel-flexible natural gas vehicles will be considerably more expensive than fuel-flexible methanol vehicles, due to the larger on-board energy storage cost (for that reason, automotive manufacturers are more willing to market a methanol vehicle than CNG vehicle); second, fuel suppliers are reluctant to invest in refueling facilities because of their high cost.

On the other hand, compared to methanol, CNG has the advantage of somewhat greater air quality benefits, lower fuel cost, a ubiquitous network of natural gas pipelines, and a large industry supporting it. The greater air quality benefit means that governments are more likely to be supportive. The low fuel cost means that once a fuel-flexible vehicle is sold, it would almost always use CNG—unlike fuel-flexible methanol vehicle owners—assuring market demand for CNG suppliers. And the existence of a large natural gas industry implies strong economic and political support for CNG.

In summary, CNG is somewhat more attractive than methanol from an environmental perspective, but faces more substantial start-up barriers. CNG will play a large role if the natural gas industry more actively supports its use, and if electric vehicles, an environmentally superior option, fail in the marketplace.

HYDROGEN AND CLEAN ELECTRICITY

Hydrogen and electric vehicles are linked here because they both are part of a potentially sustainable and very clean energy path and both could use the same clean sources of energy. Battery-powered (or roadway-powered) electric vehicles can use electricity made with solar or nuclear power (from fission or fusion reactors), and hydrogen-powered vehicles with internal combustion engines or fuel cells could use solar or nuclear power to split water to make hydrogen. This path would be followed if great emphasis is placed on reducing environmental pollution and global warming and on creating a permanently sustainable energy supply system.

Hydrogen

Hydrogen is an attractive transportation fuel in two important ways: it is the least-polluting fuel that can be used in an internal combustion engine, and it is potentially available wherever there is water and a clean source of power. The prospect of a clean, widely available transportation fuel has motivated much of the research on hydrogen fuels. The technology for cleanly producing, storing and combusting hydrogen is far from commercialization, and thus we explore a larger range of technology options in this section.

Production

Hydrogen can be produced from water or fossil fuels. Fossil fuels consist of hydrocarbon molecules that can be reformed, cracked, oxidized, or gasified to produce hydrogen. Coal is relatively abundant and could provide a low-cost feedstock for hydrogen for many decades, but if coal or other fossil fuels are to be used, it would be more attractive to convert them to liquid or gaseous fuels with a higher volumetric energy density. In addition, the conversion of fossil energy to hydrogen fuels would cause major environmental impacts and would not be a renewable energy path. Most of the hydrogen research community agrees that if hydrogen is to be used as a fuel, the most attractive source is water (Bockris, Dandapani, Cocke, and Ghoroghchian, 1985).

There are several methods for splitting water to produce hydrogen: thermal and thermochemical conversion, photolysis, and electrolysis. Electrolysis, the use of electricity to split water into hydrogen and oxygen, is the most developed method. The cost and environmental impact of producing hydrogen from water depend on the primary energy used to generate the electricity to split the water. Fossil fuels would not be used as the source of electric power because it would be cheaper and more efficient and would generate less carbon dioxide to make the hydrogen directly from the fossil fuels. Hence non-fossil feedstocks, such as solar, geothermal, wind, hydro, and nuclear energy, would be used to generate electricity for the electrolysis process. Of these, solar energy appears to have the potential to be available in the greatest quantities for the long term.

Vehicular Fuel Storage

The principal obstacle, other than costs, to using hydrogen in vehicles is hydrogen's very low volumetric energy density as a gas at ambient temperature and pressure. Hydrogen's density may be increased by storing it on board a vehicle as a gas bound with certain metals (hydrides), as a liquid in cryogenic containers, as a highly compressed gas (up to 10,000 psi) in ultra-high-pressure vessels, as a liquid hydride, and in other forms. Most research has focused on hydride and liquid hydrogen storage.

Hydride storage units, which include housings for the hydrides and the coolant systems, are very large, from 25 to over 80 gallons, and quite heavy, 250 to 1000 lbs. (Table 4-4). Barring major improvements in vehicular fuel efficiency, hydride vehicles would be limited by storage weight to a range of about 100 to 200 miles. Liquid hydrogen must be stored in double-walled, superinsulated vessels designed to minimize heat transfer and the boil-off of liquid hydrogen. Liquid hydrogen systems are much lighter and often more compact than hydride systems providing an equal range. In fact, liquid hydrogen storage is not significantly heavier than gasoline storage, on an equal-range basis, although it is about six times bulkier (Table 4-4).

In summary, all hydrogen storage systems are bulky and costly and will remain so, barring major advances that might occur with expanded research and development efforts. Hydrogen vehicles will be successfully introduced only if users are willing to accept vehicles with much larger fuel tanks and shorter ranges than other vehicles—which, as indicated later, we believe some will do, if strong incentives and social messages are given for environmentally benign and sustainable fuels.

Environmental Impacts of Hydrogen Vehicles

The attraction of hydrogen is nearly pollution-free combustion. While many undesirable compounds are emitted by gasoline and diesel fuel vehicles, or formed from their emissions, the main combustion product of hydrogen is water. Hydrogen vehicles would not produce significant amounts of CO or HCs (only small amounts from the combustion of lubricating oil), particulates, SO_x, ozone, lead, smoke, benzene, or CO_2 or other greenhouse gases (tables 4-2 and 4-3). If hydrogen is made from water using a clean power source, then hydrogen production and distribution will be pollution-free.

The only pollutant of concern in internal combustion engines (ICEs) would be NO_x, which is formed, as in all ICEs, from nitrogen taken from the air during combustion. With lean operation, and some form of combustion cooling such as exhaust gas recirculation, water injection or the use of very cold fuel (i.e., liquid hydrogen), but with no catalytic control equipment on the engine, an optimized hydrogen vehicle probably could meet the current U.S. NO_x standard, and probably have lower lifetime average NO_x emissions than a current-model catalyst-equipped gasoline vehicle (DeLuchi, 1989). And if hydrogen were used in fuel cell vehicles, there would be no NO_x emissions—in fact, no emissions of any pollutant.

The use of hydrogen made from non-fossil electricity and water is one of the most effective ways to reduce anthropogenic emissions of greenhouse gases. Highway vehicles burning hydrogen would emit essentially no CO_2 or CH_4, and because they would emit no reactive hydrocarbons (precursors to ozone formation in the troposphere), would help reduce ozone (Table 4-2).

Nuclear Versus Solar

Solar electrolytic hydrogen is environmentally and politically preferable to nuclear electrolytic hydrogen, for several reasons. First, although the nuclear power industry is developing "passively safe" reactors, such as the high-temperature gas-cooled reactor, which rely on physical laws rather than human corrective action to safely resolve emergencies (Taylor, 1989), it is not clear if the public, regulatory agencies, and financial backers will be convinced that these are safe enough to warrant a large expansion of nuclear power. Second, if nuclear power were aggressively developed, the reprocessing of spent nuclear fuel and reprocessing of plutonium for breeder reactors would circulate large amounts of weapons-grade nuclear material (Ogden and Williams, 1989). Third, and perhaps most importantly, long-term underground disposal of nuclear wastes remains environmentally controversial.

Solar power production is much less risky environmentally and politically; even concern over the amount of land devoted to photovoltaic (PV) systems may be misplaced, as it has been estimated that PV power generation (assuming 15% efficiency) requires only three times more acreage per unit of energy produced than nuclear power generation, when mining, transportation, and waste disposal are considered (Meridian, 1989).[1] In the hydrogen vehicle cost analyses below, we consider solar photovoltaic energy as the primary energy source.

---

[1] The land usage for nuclear energy estimated in Hubbard (1989) appears to have been misrepresented.

Costs

Hydrogen's environmental advantages must compensate for the very high cost of hydrogen fuel and the high cost of hydrogen storage systems. Hydrogen fuel is expensive primarily because electricity is relatively expensive (and 5% to 25% of the energy in the electricity is lost in the electrolysis process). We assume that hydrogen is produced from photovoltaic power costing between 5 and 15 cents per kWh at the generation site (Hubbard, 1989). With this assumption, Table 4-5 shows the price of gasoline that would be required to make the life-cycle cost of a gasoline and hydrogen ICE vehicle equal. In the high-cost cases, both hydride and liquid hydrogen vehicles are prohibitively expensive compared to gasoline vehicles. Even in the low-cost case, the low break-even price is about $3 per gallon; in other words, gasoline would have to sell for more than $3 per gallon for hydrogen vehicles (using hydrogen made from water with solar power) to be economically competitive. Thus, it appears that hydrogen ICE vehicles will be cost-competitive in the middle term only if the most optimistic cost projections are realized *and* the price of gasoline at least triples.

A more economical, as well as more environmentally sound, option for using hydrogen is in a fuel cell vehicle. Fuel cell vehicles are two to three times more efficient than ICE vehicles, thus reducing the cost of fuel per mile and the storage needed to attain a given range. Moreover, the electric drive system on the fuel cell vehicle would be much cheaper and longer-lasting than the power-train system for an ICE vehicle. The result, as shown in Table 4-5, is that hydrogen fuel cell vehicles can be expected to have a lower life-cycle cost than hydrogen ICE vehicles. Fuel cell vehicles represent the best long-term opportunity for hydrogen.

Opportunities for Hydrogen

The attractiveness of hydrogen vehicles hinges on technological progress in four areas. (1) In order to increase hydride vehicle range and performance, hydrides with high mass energy density, low dissociation temperature, and relatively low susceptibility to degradation by gas impurities must be found. At present, the probability of hydride vehicles achieving performance and range parity with gasoline vehicles seems low. (2) The loss of trunk space to bulky hydrogen storage systems needs to be minimized. Hydrogen storage systems are many times larger than gasoline tanks of equal range. Barring dramatic advances in technology, this disparity is not likely to change. (3) The cost of hydrogen storage must be reduced greatly. Recent developments indicate that this may be possible (see DeLuchi and Ogden, 1992). (4) Fuel cell vehicles must be successfully developed, because fuel cells represent the most promising use of hydrogen. Rapid progress is being made in reducing the cost and bulk of fuel cells (Journal of Power Sources, 1992).

The most attractive feature of hydrogen is its very low pollutant emissions, including greenhouse gases. The most fundamental barrier is cost. Therefore, if hydrogen is to be introduced as a transportation fuel, optimistic projections of the cost of hydrogen vehicles and hydrogen fuel must be realized, and a relatively high value must be placed on reducing air pollution, avoiding greenhouse warming, and reducing dependence on finite and imported energy resources.

In conclusion, while hydrogen fuel is not a near-term option, it is also not strictly an exotic, distant-future possibility. Although all hydrogen vehicles have shortcomings, none of the problems are necessarily insurmountable. With a strong research and development effort, normal technological progress, and continuing reductions in the cost of solar electricity, hydrogen vehicles could be cost-competitive on a social cost basis (taking into consideration air pollution, energy security, global warming, etc.) by the early years of the 21st century.

Electric Vehicles

A cost-effective, high-performance electric vehicle (EV), recharged quickly by solar (or perhaps nuclear) power, using widely available battery materials, would be an attractive transportation machine. Progress over the last 10 years has brought this ideal closer to reality. EVs have the potential to provide substantial air quality and petroleum conservation benefits, at comparatively low cost with acceptable performance.

Performance of EVs

Electric vehicles were commonplace in the United States at the turn of the century. However, by 1920 improvements in EV technology had lagged so far behind the development of the internal combustion engine that EVs became practically extinct (Hamilton, 1980). With the resurgence of interest in EVs in the 1960s came promises of breakthroughs that were to make EVs as economical and high-performing as internal combustion engine vehicles. But a decade later the promised EV had still not materialized.

The efforts of the past decade have not produced any dramatic break-throughs. However, over that period the technology of EV batteries and power trains has developed incrementally, and the cumulative result is substantial. For example, advances in microelectronics have resulted in low-cost, light-weight DC-to-AC inverters, which make it attractive to use AC rather than DC motors. With the improved inverters the entire AC system is cheaper, more compact, more reliable, easier to maintain, more efficient, and more adaptable to regen-erative braking than the DC systems that have been used in virtually all EVs to date. Similarly, the development of advanced batteries, particularly the high-temperature sodium/sulfur battery, has progressed to the point where success-ful commercialization does not depend on major technical breakthroughs, but on the resolution of manufacturing and quality control problems.

Advanced EVs now under development, and projected to be commercially available within a decade, are expected to offer considerably better range and performance than state-of-the art EVs of 10 years ago. The first mass-produced commercial EV is likely to be a variation of General Motors' Impact, unveiled in early 1990, and expected to be for sale around 1995. (In early 1991, GM announced that it was converting one of its factories with a capacity of about 25,000 vehicles per year to EV production.) The Impact uses advanced electric motor technology—two AC motors, one over each wheel, and a compact, effi-cient inverter—and an ultra-high-efficiency design to achieve a reported 120-mile range (at constant speed and under other artificial conditions) and per-formance equal to or better than that of comparable ICEVs.

Without sacrificing seating or cargo capacity, passenger vehicles and vans are projected to have urban ranges of about 150 miles, high top speeds and acceptable acceleration, and low energy consumption around the turn of the century. With these characteristics, EVs would be attractive as second vehicles in most multi-car households (Lunde, 1980; Horowitz and Hummon, 1987; Turrentine, Sperling, and Kurani, 1992) and as vans in most urban fleets (Berg, 1985). As personal vehicles become more specialized and expectations regard-ing multi-purpose usage of vehicles continue to diminish, EVs may become acceptable as primary commuter cars. Exotic batteries under development, such as the zinc/air and lithium/iron sulfide battery, which promise even longer ranges and faster recharging, could eventually make EVs the vehicle of choice in a world of high energy prices and heightened environmental concern.

Costs

If the most optimistic cost conditions are satisfied—high vehicle efficiency, high battery energy density, low-cost off-peak power, low initial battery cost, long battery cycle-life, long EV life, and low maintenance costs—then EVs will have life-cycle costs comparable to those of gasoline vehicles. However, under high-cost conditions, EVs will not be cost-competitive until gasoline sells for $3 to $4 per gallon (Table 4-5). If electricity is more expensive, in the range of 10 to 15 cents per kWh, the break-even price is about $4 to $5 per gallon in the high-cost case. The great difference between the high and low break-even gasoline prices is due primarily to uncertainty about the cost of batteries and the life of EVs relative to the life of internal combustion engine vehicles.

Environmental Impacts

A principal attraction of electric vehicles is the promise of improved urban air quality. If EVs use solar power, then they will be essentially nonpolluting. But even if they were to consume electricity generated in a combination of power plants using coal, natural gas, oil, hydroelectric power, nuclear power, and solar power, they would still provide a major reduction in emissions (Wang, DeLuchi, and Sperling, 1990; see Table 4-2).

Regardless of the type of power plant, fuel, and emission controls employed, EV use will practically eliminate CO and HC emissions on a per-mile basis, relative to gasoline vehicles meeting future stringent emission standards. NO_x and particulate emissions will be reduced with EV use if at least moderate controls are used. SO_x emissions will be practically eliminated if natural gas is used to generate electricity, but will increase if coal is used—by several fold, in the case of uncontrolled or moderately controlled coal steam plants. It should be noted that light-duty vehicles are now a major source of HC, CO, and NO_x emissions, but a very minor source of SO_x and particulates, and that CO and ozone are the major urban air pollution problems. Thus, a large decrease in HC, CO, and NO_x emissions from light-duty highway vehicles would have a greater impact on urban ambient air quality than would a moderate increase in SO_x emissions. As a result, regardless of the feedstock used for electricity generation, EVs will tend to improve urban air quality significantly.

The impact of EV use on greenhouse gas emissions is more mixed and more sensitive to the type of electricity feedstock used. Fossil-fuel-burning power plants emit several greenhouse gases, as well as the regulated pollutants discussed above. Table 4-3 shows the results of substituting EVs for internal combustion engine vehicles, expressed as percent change per mile in emissions of a composite greenhouse gas (CO_2 equivalents, as explained above). On a per-mile basis, the use of coal-fired power by EVs will cause a moderate increase in emissions of all greenhouse gases, relative to current emissions associated with the use of gasoline and diesel fuel. If natural gas is used, there will be a moderate decrease in emissions of greenhouse gases, mainly because of the low carbon-to-hydrogen ratio of natural gas. If EVs are powered by the marginal mix of electricity sources projected for the United States in 2000, then slightly less greenhouse gases would be emitted than will be emitted by gasoline vehicles in the year 2000. If non-fossil fuels (nuclear, solar, hydroelectric power, or biomass fuels) are used in all engines, there will be essentially zero emissions of greenhouse gases.

Opportunities for EVs

EVs will gain strong support from the electricity industry, and perhaps eventually from the automotive industry, for three reasons. First, utilities generally support the use of EVs, because they expect EVs to draw power from otherwise idle capacity and not to require the construction of new plants. Given appropriate time-of-use rates (or other load management), most recharging of EVs will be postponed until the night, when electric utilities have ample capacity available and the use of oil, which is generally a peaking fuel, is at a minimum. Studies of the impact of EV use on utility energy supply have shown consistently that utilities have sufficient capacity in place to support large numbers of electric vehicles. A detailed study of the Los Angeles area indicates that up to two million electric vehicles could be supplied with electricity without adding new generating capacity if most of the recharging took place in the evening or night (Ford, 1992).

Second, the life-cycle cost of advanced, mass-produced EVs, using cheap off-peak power, might be low enough to induce some fleet operators and home owners to purchase those vehicles. Third, vehicle sales will not be hindered initially as much as methanol and CNG vehicles by the absence of a fuel distribution network, because one already is in place. Electricity is available virtually everywhere, and most homes and businesses can set up an EV charging station for well under $1,000 (Hamilton, 1988; Nesbitt, Kurani, and DeLuchi, 1992). These relatively small cost and start-up barriers (the "chicken-and-egg" problem) means that the market penetration of EVs can proceed, to a point, largely by market forces.

The degree of market penetration by EVs will depend initially on their range, performance, and life-cycle cost. In the near future, EVs will be attractive in some urban fleets; as the technology improves and vehicles are produced in large quantities, EVs may be attractive as commuter vehicles. However, even if advanced EVs prove to be as high-performing and economical as can be hoped, and are favored by public policy for their environmental benefits, there still will be one significant obstacle to widespread consumer acceptance: the long recharging time. If it takes 8 hours to recharge an EV, most households will want at least one non-electric vehicle, and EVs will be limited to the role of second car in some multi-car, home-owning households. However, if EVs can be charged in under 30 minutes, they may be able to displace gasoline vehicles in many more applications, and gain a large share of the vehicle market; they may be suitable for all applications except those requiring more power than even advanced batteries can provide.

There are several ways of quickly recharging EVs, including swapping discharged batteries for previously fully charged ones, using mechanically rechargeable batteries, and using ultra-high-current recharging. None of these methods has been adequately demonstrated, however, and all are likely to be expensive. The successful development of fuel cell electric vehicles, though, could obviate the problem of recharging time because fuel cells can be "recharged" (refueled) in the same way as a methanol or hydrogen ICE vehicle.

Continued improvements in advanced batteries and fuel cells and a means for quickly recharging EVs, would make the EV a competitive alternative to internal combustion vehicles. The combination of large environmental benefits and potentially low private cost in the near term, and the prospect of a pollution-free feedstock in the long run, may well make EVs the option with the lowest social cost. In the meantime, though, EVs may be economical, on a private-cost basis, in some applications in the very near future.

In summary, EVs and hydrogen vehicles require substantial improvements before they become attractive as the dominant transportation technology. For that to happen, research and development investments must be expanded greatly. A clean electricity and hydrogen path will come into being in a timely manner only if society places much greater emphasis on the need to reduce air pollution and slow the greenhouse effect.

CONCLUSIONS AND RECOMMENDATIONS

No analytical basis exists for definitively determining which fuel is superior and when it should be introduced. The price of petroleum cannot be predicted, and many of the costs and benefits of alternative fuels are difficult to quantify.

Moreover, different regions and groups of people place differing values on the important (nonmarket) concerns: energy security, air quality, global warming, and the ease and convenience of a transition. In short, different beliefs and different values, and familiarity with different facts, lead individuals and organizations to different conclusions about the most desirable path.

There is no one optimal choice; the era of one (or two) uniform transportation fuels is nearing an end. This prospective multiplicity of fuel options presents a challenge for business and government. Because many of the benefits resulting from initial alternative fuel investments do not accrue to the private-sector supplier of the fuel, government must take much of the initiative. But which fuels should it choose and how fast should it introduce them?

If concerns for self-sufficiency and energy independence dominate, then the United States would favor fuels from biomass, coal, oil shale, and domestic natural gas, as well as electric vehicles.

If economic efficiency, measured by conventional market indicators, is the dominating value, then hydrogen ICE vehicles would be discarded as an option, while electric and hydrogen fuel cell vehicles would be competitive in some applications if optimistic battery and fuel cell cost and performance goals were met. For the larger passenger and heavy-duty vehicle markets, natural gas vehicles probably would be favored, as would methanol if low-cost methanol production estimates prove accurate.

If the overriding objective is to make the transition to new fuels with as little disruption as possible, then reformulated gasoline (and diesel) fuel should be pursued in the very near term, followed by petroleum-like fuels and alcohol fuels. A transition to petroleum-like fuels would require no significant changes to vehicles and fuel distribution systems, but would be the least attractive environmentally and among the more expensive options. A transition to methanol would be more difficult than a transition to petroleum-like fuels, requiring modifications to vehicles, storage tanks, and delivery systems, but it would be less difficult than a transition to gaseous fuels. A transition to EVs would be relatively easy from an infrastructure standpoint, but not from a consumer acceptance perspective because of the weight and low energy density of batteries and the long recharging time.

If environmental quality and sustainability takes precedence, then hydrogen, electric, and fuel cell vehicles, using clean and renewable energy (probably solar power), would be preferred. Methanol and natural gas vehicles, regardless of the feedstocks, would be deployed as transitional options only, if at all.

If the most important concern is to avoid a greenhouse warming, battery- or fuel cell-powered electric vehicles, relying ultimately on non-fossil energy, are the best choice, because they completely eliminate fuel-cycle emissions of greenhouse gases. ICE vehicles using hydrogen made from water with non-fossil power would also emit only negligible amounts of greenhouse gases, but hydrogen ICE vehicles are not likely to be commercially available as soon as electric vehicles. ICEVs using methanol or gas derived from woody biomass would emit only small amounts of greenhouse gases, but the biomass resource base is limited, the use of these biomass fuels is much more polluting than the use of clean power by EVs, and biomass cultivation demands careful soil management.

Other values and goals could and should play instrumental roles—equity and distribution of power and wealth, growth versus stability, free enterprise, individual initiative, and public health—but the issues discussed here of environmental quality, greenhouse effects, sunk investment, compatibility, and energy security have come to dominate the public debate.

The rational analyst despairs of making choices in this context of competing goals and values, and with incomplete knowledge and limited foresight. Nonetheless, choices must be made. Based on a belief that a transition to alternative fuels is important but not extremely urgent at this time and that pollution and environmental damage should weigh heavily in decision-making, we suggest the following:

In the near term, CNG and EVs should be aggressively promoted so as to expand their use in market niches where they are economically attractive. The cost of electric vehicles, in particular, will drop sharply as technology development efforts are accelerated and economies of scale are realized. Natural gas and electric vehicles are the only options that (in the best case) can compete with gasoline when oil is in the range of $20 per barrel. They should especially be supported in areas with CO and ozone problems.

In the near term, in most regions, government should give priority to the introduction of electric, natural gas and methanol fuel—in that order. Methanol offers only limited air quality and energy security benefits, and no greenhouse benefits and, unlike CNG and EVs, is not attractive in any market niches. Perhaps methanol is best treated as a transitional "filler," along with natural gas vehicles, after CNG and EV niche markets are saturated and oil prices rise on a sustained basis to perhaps $30 per barrel.

The long-term and possibly permanent transportation fuels will probably be a mix of electricity, hydrogen, and biomass fuels used in battery-powered and fuel cell vehicles. These fuels produce much fewer greenhouse gas emissions, are more environmentally benign than other options, and can be supplied on a permanently sustainable basis. Since they provide large social benefits that are mostly ignored by the marketplace, they merit the most attention from government.

The worldwide petroleum-based transportation energy system now in place has served us well. But it will not serve us well in the future. The skewed distribution of petroleum reserves, the rapid worldwide growth of motor vehicles, the increasing concentration of greenhouse gases in the atmosphere, and persistent local air pollution problems, together are creating ever greater political and environmental stresses. Government policies and initiatives should be restructured so as to provide a framework for guiding private investments and consumer choices toward the options that can really make a difference. New transportation energy options should not be treated as second- best alternatives to petroleum but as opportunities to make a better world.

REFERENCES

Acurex (1982). "Clean Coal Fuels: Alternate Fuel Strategies for Stationary and Mobil Engines, Executive Summary." Report P500-82-020. Sacramento: California Energy Commission.

Alson, J., J. Adler, and T. Baines (1989). "Motor Vehicle Emission Characteristics and Air Quality Impacts of Methanol and Compressed Natural Gas." In D. Sperling, ed., *Alternative Transportation Fuels, An Environmental and Energy Solution*. Westport, Conn.: Quorum Books/Greenwood Press.

American Gas Association (1985). "The Gas Energy Supply Outlook Through 2010." Arlington, Va.

Bechtold, R. L. (1987). "Compendium of Significant Events in the Recent Development of Alcohol Fuels in the United States." Oak Ridge National Laboratory Report ORNL/Sub/85-22007/1. Springfield, Va.: National Technical Information Service.

Berg, L. (member of executive committee of South Coast Air Quality Management District) (1984). Testimony to U.S. Congress, House of Representatives, Committee on Energy and Commerce, Subcommittee on Fossil and Synthetic Fuels, Methanol as Transportation Fuel, 98th Cong., 2d sess. April 4 and 25, 1984. 98-145, p. 126. Washington, D.C.: Government Printing Office.

Berg, M. (1985). "The Potential Market for Electric Vehicles: Results from a National Survey of Commercial Fleet Operators." *Transportation Research Record*, Vol. 1049, pp. 70-78.

Bockris, J., B. Dandapani, D. Cocke and J. Ghoroghchian (1985). "On the splitting of water." *International Journal of Hydrogen Energy*, Vol. 10, pp. 179-201.

Boekhaus (1990). "Reformulated Gasoline for Clean Air: an ARCO Assessment." Presented at 2nd Biennial UC Davis Conference on Alternative Transportation Fuels, July 11-13, Davis, California.

California Advisory Board on Air Quality and Fuels (1989). Report to the California Legislature, Vol. 1, Executive Summary. October. Sacramento.

California Air Resources Board (1988). California Air Quality Data: Summary of 1987 Air Quality Data, Vol. XIX. Sacramento.

California Air Resources Board (1990). Proposed Regulations for Low-Emission Vehicles and Clean Fuels. Staff Report. August 13. Sacramento.

California Energy Commission (1986a). Energy Development Report, P500-86-001. Sacramento.

California Energy Commission (1986b). Fuels Report, P300-86-007. Sacramento.

California Energy Commission (1987). Fuels Report, P300-87-016. Sacramento.

California Energy Commission (1991). Fuels Report. Sacramento.

Cann, J. H. (1988). Presentation by B.C. Hydro of Canada to Gas Symp. April 19. Washington, D.C.

Carter, W. P. L., R. Atkinson, W. D. Long, L. L. N. Parker, and M. C. Dodd (1986). "Effects of Methanol Fuel Substitution on Multi-Day Air Pollution Episodes." Air Resources Board report ARB-86. Riverside: University of California, Statewide Air Pollution Research Center.

Chadwick, M. J., N. H. Highton, and N. Lindman, eds. (1987). *Environmental Impacts of Coal Mining and Utilization*. Oxford: Pergamom.

DeLuchi, M. A. (1989). "Hydrogen Vehicles: an Evaluation of Fuel Storage, Performance, Safety, Environmental Impacts, and Cost." *International Journal of Hydrogen Energy*, Vol. 14, pp. 81–130.

DeLuchi, M. A. (1991). "State-of-the-art Assessment of Emissions of Greenhouse Gases from the Use of Fossil and Nonfossil Fuels, with Emphasis on Alternative Transportation Fuels." To be published. Argonne, Ill.: Argonne National Laboratory, Center for Transportation Research.

DeLuchi, M. A., R. A. Johnston, and D. Sperling (1988). "Methanol vs. Natural Gas Vehicles: A Comparison of Resource Supply, Performance, Emissions, Fuel Storage, Safety, Costs, and Transitions." SAE Report 881656. Prepared for International Fuels and Lubricants Meeting, Portland, Oregon, October 10–13. Warrendale, Pa.: Society of Automotive Engineers, Inc.

DeLuchi, M. A, E. D. Larson, and R. H. Williams (1991). "Hydrogen and Methanol: Production from Biomass and Use in Fuel Cells and Internal Combustion Engine Vehicles." Report No. 263. Princeton, New Jersey: Princeton University, Center for Energy and Environmental Studies.

DeLuchi, M. A. and J. Ogden (1992). "Solar Hydrogen Transportation Fuel." Report. Princeton, New Jersey: Princeton University, Center for Energy and Environmental Studies.

DeLuchi, M. A., D. Sperling, and R. A. Johnston (1987). "A Comparative Analysis of Future Transportation Fuels."Report UCB-ITS-RR-87-13, p. 364. Berkeley: University of California, Institute of Transportation Studies.

DeLuchi, M. A., Q. Wang, and D. Sperling (1989). "Electric Vehicles: Performance, Lifecycle Costs, Emissions, and Recharging Requirements." *Transportation Research*, Vol. 23A, pp. 255–278.

Difiglio, C. (1989). Results from U.S. Department of Energy Alternative Fuels Trade Model. Personal communication.

Ford, A. (1992). "The Impact of Electric Vehicles on the Southern California Edison System." Report prepared for California Institute for Energy Efficiency. Berkeley, Calif.: Lawrence Berkeley Laboratory.

Geller, H. (1985). "Ethanol fuel from sugar cane in Brazil." *Annual Review of Energy*, Vol. 10, pp. 135–164.

Gray, C. and J. Alson (1985). "Moving America to Methanol: A Plan to Replace Oil Imports, Reduce Acid Rain, and Revitalize our Domestic Economy." Ann Arbor: University of Michigan Press.

Gray, C. and J. Alson (1989). "The Case for Methanol." *Scientific American*, Vol. 261, No. 5, pp. 108–114.

Hamilton, W. (1980). "Electric Automobiles." New York: McGraw-Hill.

Hamilton, W. (1988). "Electric and Hybrid Vehicles, Technical Background Report for the DOE Flexible and Alternative Fuels Study." Report prepared for U.S. Deptartment of Energy.

Harris, J. N., A. G. Russell, and J. B. Milford (1988). "Air Quality Implications of Methanol Fuel Utilization." Report SAE 881198. Warrendale, Pa.: Society of Automotive Engineers.

Harvey, G. (1990). Unpublished analysis for Metropolitan Transportation Commission. March. Oakland, Calif.

Holcomb, M. C., S. Floyd, and S. L. Cagle (1987). "Transportation Energy Data Book," Edition 9. Oak Ridge National Laboratory Report ORNL-6325. Springfield, Va.: National Technical Information Service.

Horowitz, A. and N. P. Hummon (1987). "Exploring Potential Electric Vehicle Utilization: a Computer Simulation." *Transporation Research*, Vol. 21A, pp. 17–26.

Hubbard, H. M. (1989). "Photovoltaics today and tomorrow." *Science*, Vol. 244, pp. 297–304.

Jet Propulsion Lab (1983). "California Methanol Assessment #83-14." Pasadena, Calif.

Journal of Power Sources (1992). Special issue: "Progress in Fuel Cell Commercialization," Proc. 2nd Grove Fuel Cell Symposium. *Journal of Power Sources*, Vol. 37, No. 1-2.

Kant, F., A. Cohn, A. Cunningham, M. Farmer, and W. Herbst (1974). "Feasibility Study of Alternative Fuels for Automotive Transport." Report PB-23 5580, 3 Vols. Prepared by Exxon Research and Engineering Co. for the U.S. Environmental Protection Agency. Springfield, Va.: National Technical Information Service.

Krupnick, A. J. and M. A. Walls (1992). "The Cost-Effectiveness of Methanol for Re-ducing Motor Vehicle Emissions and Urban Ozone Levels." *Journal of Policy Analysis and Management* to be published.

Larson, W. (1979). "Crop Residues: Energy Production or Erosion Control." *J. Soil and Water Conservation*, Vol. 34, pp. 74–76.

Lumpkin, R. E. (1988). "Recent progress in the direct liquefaction of coal." *Science*, Vol. 239, pp. 873–77.

Lunde, L. G. (1980). "Some Potential Impacts on Travel of Alternative Assumptions on the State of Electric Vehicle Technology." *Transport Policy and Decision Making*, Vol. 1, pp. 361–377.

Lynd, L. R., J. H. Cushman, R. J. Nichols, and C. E. Wyman (1991). "Fuel Ethanol from Cellulosic Biomass." *Science*, Vol. 251, pp. 1318–1323.

McNutt, B. and E. E. Ecklund (1986). "Is There a Government Role in Methanol Market Development?" Report SAE 861571. Warrendale, Pa.: Society of Automotive Engineers.

Meridian Corp (1989). "Energy Systems Emissions and Material Requirements." Re-port prepared for the U.S. Department of Energy.

Murrell, J. D. and G. K. Piotrowski (1987). "Fuel Economy and Emissions of a Toyota T-LCS-M Methanol Prototype Vehicle." Report SAE 871090. Warrendale, Pa.: Society of Automotive Engineers.

National Research Council (1990). "Fuels to Drive our Future." Washington, D.C.: National Academy Press.

Nesbitt, K., K. A. Kurani, and M. A. DeLuchi (1992). "Home Recharging and the Near Term Market for Electric Vehicles: A Constraints Analysis." *Transportation Research Record*, to be published.

Nichols, R. J. and J. M. Norbeck (1985). "Assessment of emissions from methanol-fueled vehicles: implications for ozone air quality." Presented at 78th Annual Meeting of the Air Pollution Control Association, Detroit.

Office of Technology Assessment (1980a). "Energy from Biological Processes." Vol. II, p. 173. Washington, D.C.: Government Printing Office.

Office of Technology Assessment (1980b). "An Assessment of Oil Shale Technology." Washington D.C.: Government Printing Office.

Office of Technology Assessment (1989). "Catching Our Breath: Next Steps for Reducing Urban Ozone." OTA-O-413. July. Washington, D.C.: Government Printing Office.

Office of Technology Assessment (1990). "Replacing Gasoline: Alternative Fuels for Light-Duty Vehicles." Washington, D.C.: Government Printing Office.

Ogden, J. M. and R. H. Williams (1989). "Hydrogen and the Revolution in Amorphous Silicon Solar Cell Technology." Report No. 231. February 15. Submitted to *International Journal of Hydrogen Energy*. Princeton, New Jersey: Princeton University, Center for Energy and Environmental Studies.

Pace Co. Consultants and Engineers (1981). "Comparative Analysis of Coal Gasification and Liquefaction." Prepared for Acurex Corp. and the California Energy Commission. Denver, Colo.

Perry, H. and H. Landsberg (1981). "Factors in the Development of a Major U.S. Synthetic Fuels Industry." *Annual Review of Energy*, Vol. 6, pp. 233-266.

Pimentel, D. (1981). "Biomass Energy from Crop and Forest Residues." *Science*, Vol. 212, p. 1115.

Portney, P. R., D. H. Harrison, Jr., A. J. Krupnick, and H. Dowlatabadi (1989). "To Live and Breathe in L.A." *Issues in Science and Technology*, Vol. V, No. 3, pp. 68-73.

Purdue University (1981). "Transportation Energy Futures: Paths of Transition." West Lafayette, Ind. Partial results were published in Brinkley, J. K., W. Tyner, and M. Mathews, "Evaluating alternative energy policies." *The Energy Journal*, Vol. 4, pp. 91-104.

Russell, A. G. (1987). "Quantitative Estimates of Air Quality." Report SAE 872054. Warrendale, Pa.: Society of Automotive Engineers.

Santini, D. J. and J. J. Schiavone (1988). "Technical Problems and Policy Issues Associated With the 1991 Bus Emission Standards." *Transportation Research Record*, Vol. 1164, pp. 5-14.

Sathaye, J., B. Atkinson, and S. Myers (1989). "Promoting Alternative Transportation Fuels: the Role of Government in New Zealand, Brazil, and Canada." In D. Sperling, ed., *Alternative Transportation Fuels: An Energy and Environmental Solution*. Westport, Conn.: Quorum Books/Greenwood Press.

Science (1990). "New Greenhouse Report Puts Down Dissenters." Vol. 249, p. 481.

Sierra Research, Inc. (1988a). "The Feasibility and Costs of More Stringent Mobile Source Emissions Controls." Report prepared for the U.S. Congress, Office of Technology Assessment. January 20. Sacramento.

Sierra Research, Inc. (1988b). "Potential Emissions and Air Quality Effects of Alternative Fuels." Report SR88-11-02. Sacramento.

Smith, K. D., D. W. Fong, D. S. Kondoleon, and L. S. Sullivan (1984). Proc. International Alcohol Fuel Symp. on Alcohol Fuel Tech., Ottawa, Canada, pp. 2-373 to 2-383.

Soil Conservation Service (1978). "National Erosion Inventory Estimate." Washington D.C.: U.S. Department of Agriculture.

Sperling, D. (1987). "Brazil, Ethanol and the Process of System Change." *Energy*, Vol. 12, pp. 11–23.

Sperling, D. (1988). *New Transportation Fuels: A Strategic Approach to Technological Change*. Berkeley: University of California Press.

Sperling, D. and M. A. DeLuchi (1989). "Is methanol the transportation fuel of the future?" *Energy*, Vol. 14, pp. 469–489.

SRI International (1976). "Synthetic Liquid Fuels Development: Assessment of Critical Factors." U.S. Energy Research and Development Administration Report ERDA 76-129/1. Washington, D.C.: Government Printing Office.

Systems Application, Inc. (1984). "The Impact of Alcohol Fuels on Urban Air Pollution: Methanol Photochemistry Study." U.S. Department of Energy report DOE/CE/50036-1. Springfield, Va.: National Technical Information Service.

Tanner, R., A. H. Miguel, J. B. de Andrade, J. S. Gaffney, and G. E. Streit (1988). "Atmospheric Chemistry of Aldehydes: Enhanced Peroxyacetyl Nitrate Formation from Ethanol Fueled Vehicular Emissions." *Environmental Science and Technology*, Vol. 22, pp. 1026–1034.

Taylor, J. J. (1989). "Improved and Safer Nuclear Power." *Science*, Vol. 244, pp. 318–325.

Tesche, T. W. (1984). "Photochemical Dispersion Modeling: Review of Model Concepts and Applications Studies." *Environment International*, Vol. 9, pp. 465–489.

Three-Agency Methanol Task Force (1986). Report, prepared jointly by California Air Resources Board, California Energy Commission, and South Coast Air Quality Management District. Sacramento.

Trindade, S. and A. V. de Carvalho (1989). "Transportation fuels policy issues and options: the case of ethanol fuels in Brazil." In D. Sperling, ed., *Alternative Transportation Fuels: An Environmental and Energy Solution*, pp. 163–186. Westport, Conn.: Quorum Books/Greenwood Press.

Turrentine, T. D. Sperling, and K. A. Kurani (1992). "Market Potential for Electric and Natural Gas Vehicles." Prepared for the California Insitute for Energy Efficiency. Berkeley, Calif.: Lawrence Berkeley Laboratory.

Ullman, T. L. and C. T. Hare (1982). "Emission Characterization of Spark-Ignited, Heavy-Duty, Direct-Injected Methanol Engine." U.S. Environmental Protection Agency Report EPA 460/3-82-003. Ann Arbor, Mich.

U.S. Department of Energy (1987). "Energy Security." Report DOE/S-0057. Washington, D.C.

U.S. Department of Energy (1988). "Assessment of Costs and Benefits of Flexible and Alternative Fuel Use in the U.S. Transportation Sector." Washington, D.C.: Government Printing Office.

U.S. Department of Energy (1990). "Vehicle and Fuel Distribution Requirements; Technical Report Four, Assessment of Costs and Benefits of Flexible and Alternative Fuel Use in the U.S. Transportation Sector." Washington, D.C.: Government Printing Office.

U.S. Department of Energy, Energy Information Administration (1992). "Annual Energy Outlook—with Projections to 2010." DOE/EIA-0383(92). March. Washington, D.C.: Government Printing Office.

U.S. General Accounting Office (1990). "Gasoline Marketing: Uncertainties Surround Reformulated Gasoline as a Motor Fuel." Report GAO/RCED-90-153. Washington, D.C.: Government Printing Office.

U.S. Synthetic Fuels Corp (1985). "Comprehensive Strategy Report." Appendices and Final Report. Washington, D.C.: Government Printing Office.

Wang, Q., M. A. DeLuchi and D. Sperling (1990). "Emission impacts of electric vehicles." *Journal of Air and Waste Management Association*, Vol. 40, pp. 1275-1284.

Wolk, R. and N. Holt (1988). "The Environmental Performance of Integrated Gasification Combined Cycle (IGCC) Systems." Palo Alto, Calif.: Electric Power Research Institute.

Wright, J. D. (1988). "Ethanol from biomass by enzymatic hydrolysis." *Chemical Engineering Progress*, Vol. 84, No. 8, pp. 62-74.

VEHICLE EMISSIONS, URBAN SMOG, AND CLEAN AIR POLICY

5

Alan J. Krupnick

INTRODUCTION

Currently, some 96 mostly urban areas violate the National Ambient Air Quality Standards (NAAQS) for ozone and 41 violate the standard for carbon monoxide (CO) (U.S. Environmental Protection Agency, 1991a). If U.S. cities are to meet these standards, major reductions in emissions of CO and the ozone precursors—nitrogen oxides (NO_x) and reactive volatile organic compounds (VOCs)[1]—will be necessary. As gasoline vehicles (GVs) contribute most of the CO emissions and a large share of VOCs and NO_x, reductions in their emissions must be a component of strategies to attain the ambient standards.

This paper considers the contribution of vehicular emissions to total emissions and then offers a variety of hypotheses for why the impressive reductions in emissions of new vehicles have not resulted in attainment of the NAAQS. The focus is on attaining the ozone standard, because the least progress has

Alan J. Krupnick is a Senior Fellow with Resources for the Future, Quality of the Environment Division. The author would like to acknowledge the helpful comments of Richard Gilbert and Daniel Sperling on earlier versions of this paper.

[1] These compounds are more generally referred to as hydrocarbons (HC), which include VOCs and non-reactive hydrocarbons such as methane. The reactive portions are also called non-methane hydrocarbons (NMHC) and reactive organic gases (ROGs). These terms will be used interchangeably in this paper.

been made, nor is expected to occur, on this pollutant.[2] Based on this discussion, the cost-effectiveness of alternative policies for reducing VOC emissions and meeting ozone standards is considered.

EMISSIONS FROM VEHICLES

Just how much vehicles can contribute to emissions reductions depends in part on their current share of emissions. Considering estimates of national emissions in 1989, highway vehicles, the dominant source of transportation emissions, contributed 54% of the CO, 28% of the VOCs,[3] and 30% of the NO_x (U.S. Environmental Protection Agency, 1991b). Gasoline vehicle emissions were responsible for most of the CO and VOCs emitted by highway vehicles, but diesel vehicles contributed about 40% of the NO_x (Figure 5-1).

However, these national figures underestimate the share of vehicle emissions within nonattainment areas, particularly for NO_x, which is produced in large quantities from power plants, many of which are located in rural areas. Estimates of emissions in nonattainment areas are harder to come by. But the Office of Technology Assessment (1989) found that (highway) vehicles contributed 45% of the VOCs in "ozone nonattainment" areas in 1985. This share varied across U.S. urban areas from a low of 30% to a high, in Los Angeles, of about 66%. In the major urban areas of California, the California Air Resources Board (CARB) (Burmich, 1989) estimates that cars and trucks contribute 43% of the ROGs, and 57% of the NO_x.

Even these estimates may downplay the importance of vehicular emissions if the yearly emissions estimates are unrepresentative of days on which the NAAQS are violated. In general, highway travel is more concentrated in the daylight hours than emissions from other sources. As ozone is created in the presence of sunlight, vehicles may be responsible for a larger portion of ozone creation than is implied by a daily average estimate of emissions. The continued large share of emissions contributed by vehicles is more a reflection of continued reductions in emissions from all sources than a lack of progress on

---

[2] As CO emissions reductions appear to be successfully reducing CO concentrations, this discussion refers only to the link between ozone precursor emissions and ozone concentrations.

[3] Indeed, MOBILE4 has come under much criticism for underestimating VOC emissions (Stedman 1989; Pierson, Gertler, and Bradow, 1990; Guensler and Geraghty, 1991).

vehicular emissions. Reductions in vehicular emissions (on a per mile basis) in new cars have been impressive. Federal tailpipe emissions standards were first set for vehicles in 1968, were gradually tightened until 1981,[4] and then were tightened further in the 1990 Clean Air Act Amendments (CAAA).

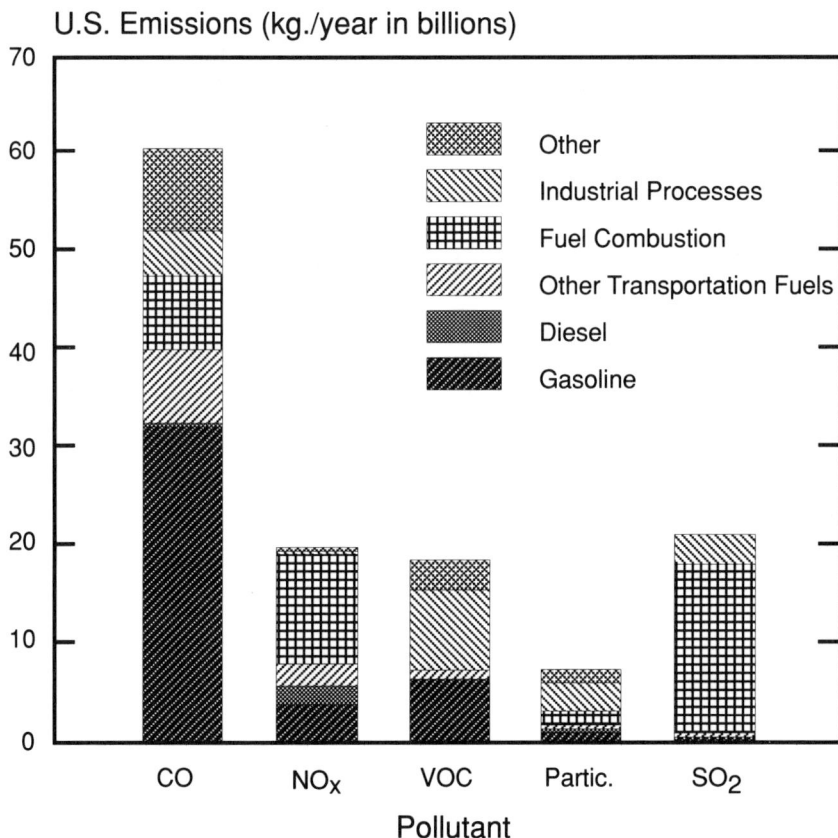

FIGURE 5-1   Annual U.S. Emissions by Source, 1989. SOURCE: U.S. Environmental Protection Agency (1991b), pp. 18–22.

---

[4]   The 1981 standards represent a decline in uncontrolled emissions per mile of 96% for HC and CO and 75% for NO_X.

EFFECTS ON AIR QUALITY

The reductions in VOC and NO_x emissions from vehicles and other sources have not reduced concentrations of ozone in urban areas. The composite average of the second highest 1-hour ozone concentration taken during a day shows basically no trend over the 1980–89 period. Because summer 1989 was particularly cool, daily ozone exceedences were 53% below 1980 levels; however, in 1988, when the summer was hot, exceedences were only 10% below 1980 levels (U.S. Environmental Protection Agency, 1991a). Thus, transitory changes in weather, rather than emissions reductions, appear to have resulted in a drop in the population living in nonattainment areas from some 112 million people in 101 urban areas in 1988 to 66.7 million people in 96 areas in 1989.

WHY DOES NONATTAINMENT PERSIST?

The above discussion raises two questions: Why have reductions in tailpipe standards failed to reduce emissions commensurately, and why have the substantial reductions in emissions that have occurred failed to improve ozone levels in urban areas? If these questions can be answered, policies to address vehicular emission could be improved.To examine the possible explanations, the following equation and identity will prove helpful:

$$C_j = f(P_i, B_i, M) \qquad\qquad [1]$$

$$P_i \equiv \sum_{t,k} \text{emissions/mile} \times \text{MPG} \times \text{gallons per trip} \qquad [2]$$
$$\times \text{ trips per vehicle} \times \text{\# vehicles}$$

For equation [1], C_j is the daily maximum concentration of ozone, P_i is emissions of pollutant i per period (i = VOCs or NO_x), B_i is background (ambient) concentrations of VOCs and NOx, and M represents a number of climate variables such as temperature, sunlight, humidity, etc.

For identity [2], emissions of a pollutant i are summed over t vehicle types (e.g., t = gasoline autos, diesel buses) and k age classes. The variables in [2] each may be differentiated by t and k. The emissions-per-mile variable is itself dependent on the emissions standards, on actual driving conditions, on characteristics of the fuel being used, on vehicle inspection and maintenance programs, and on deterioration of and the degree of tampering with emissions control equipment and other components of the vehicle.

To convert emissions per mile into total emissions, one can multiply the former by an estimate of vehicle miles traveled (VMTs). However, the use of a VMT measure obscures factors that have policy implications. Hence, in [2] this term is disaggregated into four terms, which have more behavioral content. The first, miles per gallon, depends on the fuel efficiency (CAFE) standards and fuel type, as well as driving habits and vehicle condition. Gallons per trip is inserted to highlight the effects of congestion on mileage. Trips per vehicle is delineated to highlight the type of trips taken [even this term can be further disaggregated into trips per person multiplied by occupancy rates (persons per vehicle)]. Number of vehicles is included to address vehicle ownership decisions.

Determinants of Ambient Ozone Concentrations

Nonattainment may be persisting in some areas simply because policy generally has been focusing on reducing VOCs, rather than NO_x, even though the former may be unproductive in reducing ozone in some areas (National Research Council, 1991). The effect of VOC and NO_x emissions on ambient ozone in a region is highly nonlinear and dependent on the ratio of ambient VOCs to NO_x on any given day. In areas with high ratios, reductions in NO_x are likely to be more productive at reducing peak ozone concentrations than reducing VOCs. The converse is true for areas with low ratios. Table 5-1 (Dowlatabadi, Krupnick, and Russell, 1990, from Table 5.2) provides an indication of the range of ozone sensitivities to precursor emission reductions for different cities. It can be seen that NO_x reductions are more productive in Atlanta but less productive in the other cities listed, while the sensitivity of ozone to reductions in either pollutant across cities varies widely, particularly for NO_x.

Indeed, the sensitivity of ozone to NO_x and VOC emissions reductions may vary temporally, even within the same city. Milford, Russell, and McRae (1989) show, for Los Angeles, that VOC and NO_x reductions are about equally productive up to ozone levels of 0.17 ppm (the ozone standard is 0.12 ppm for the daily 1-hour peak concentration) but that reductions below 0.15 ppm can only be obtained by reducing NO_x. Temperature, winds, and mixing heights can also affect these relationships.

TABLE 5-1   Normalized Sensitivities of Peak Ozone to Pollutant.

| City | — Reductions — | |
	ROG	NO_x
Atlanta	0.50	0.54
Boston	0.70	0.31
St. Louis	0.54	0.19
New York	0.64	-0.24
Average	0.59[a]/0.54[b]	0.18
Los Angeles[c]	0.51/1.22	-0.20/0.43

SOURCE: Dowlatabadi, Krupnick, and Russell (1990).
[a] Average of Atlanta, Boston, St. Louis and New York.
[b] As calculated from Chang, Rudy, Kuntasal, and Gorse (1989), for the average of the 20 cities.
[c] Sensitivity of ozone exposure based on Russell, Harris, St. Pierre, and Milford (1989). The value preceding the slash is total exposure to ozone, and the value after it is the exposure to ozone $\geq$ 0.12 ppm. Note the change in sensitivity to NO_x.

This differential sensitivity of ozone to changes in emissions has important implications for the cost-effectiveness of mobile source pollution control strategies. It suggests, first, that the cost-effectiveness of meeting the national ozone standard will vary widely across cities and, second, that the focus on VOC controls in the Clean Air Act and its amendments, may be inappropriate for some cities. Finally, it suggests that policies need to be fine-tuned and flexible, as an emphasis on controlling VOCs today may need to give way to tomorrow's emphasis on NO_x controls. The overarching implication for policy is that strategies designed in light of local conditions are likely to be more effective at reducing ozone concentrations than those mandated by the federal government.

Determinants of Vehicular Emissions

Emissions per mile. In spite of tight emissions standards on new cars and the gradual scrapping of old, dirty cars, reductions in estimated actual emissions per mile have been smaller than might be expected. There are several important reasons for this.

(1) Estimates of actual emissions per mile exceed the vehicle emissions standards by a large margin. The standards are required to be met by a vehicle when it reaches 50,000 miles; but in practice, vehicles need only be certified to meet these standards based on a test procedure applied to a sample of vehicles,

only a few of which are actually driven 50,000 miles. In fact, because of tampering with emissions control devices, use of leaded fuel in cars with catalysts, poor maintenance, and an emissions test procedure that may not be representative of urban driving conditions, in-use emissions per mile, according to EPA, exceed the standards. For instance, EPA's MOBILE4 model (Shih, 1990) predicts in-use exhaust HC emissions for 1985+ model-year vehicles of 0.60 gram per mile versus the standard of 0.41 gram per mile. For 1980-model-year vehicles actual 50,000 mile emissions are much larger (1.3 grams per mile) compared to the same standard,[5] and, as noted above, MOBILE4 may systematically underestimate VOCs.[6]

(2) The emissions standards have been primarily directed at tailpipe emissions, to such an extent that other sources of vehicular VOC emissions now dominate. For a new car, 41% of total VOCs come from the tailpipe, the rest coming from evaporative (16%), refueling (11%), and running loss (31%) emissions (U.S. Environmental Protection Agency, 1989).[7]

(3) Changes in the composition of gasoline have led to increases in its volatility, which have raised VOC emissions. The volatility of gasoline [as measured by Reid Vapor Pressure (RVP)] has risen from 9 psi (when emissions standards were originally set) to 11 psi, which prevailed in the 1985–88 period (Office of Technology Assessment, 1989). According to MOBILE4, on a hot summer day, this increase in RVP can cause over a doubling of hot soak and running loss emissions (Shih, 1990).[8]

(4) Because people are holding onto their vehicles for longer periods of time (see below), average emissions per mile increase, *ceteris paribus*. The

---

[5] The above discussion of emissions assumed a specific driving cycle. MOBILE4 produces estimates of emissions based on a given percentage of cold start, hot start, and stabilized VMTs and a driving cycle with average speed of 19.6 MPH, the urban average.

[6] As long as the model tracks progress in emission *changes* reasonably well, its failure to estimate average emissions accurately cannot explain lack of progress on reducing ozone concentrations.

[7] Evaporative emissions include emissions from a gas tank as ambient temperature rises during the day (diurnal emissions) and from a hot engine after it is turned off (hot soak emission). Running loss emissions are emissions that occur while the vehicle is running but do not come from the tailpipe; refueling emissions occur when gasoline is added to the tank.

[8] Note, however, that as RVP has increased, there has been an offsetting reduction in the proportion of hydrocarbons with the highest ozone-forming potential (Williams, Lipari, and Potter, 1989).

average age of passenger vehicles has lengthened from 5.7 years in 1974 to 7.6 years in 1989. Moreover, the age distribution of the fleet has significantly flattened since 1975. In 1989, 31% of vehicles were 10 years or older and 8% of these were 16 years or older; in 1980, only 20.5% of vehicles were 10 years or older and 3.4% of these were 16 years or older. A similar aging can be seen in the truck fleet, which is dominated by light-duty trucks facing the same emissions standards as passenger vehicles (Motor Vehicle Manufacturers Association, 1990).

Why is this aging of the fleet so important? To begin, a pre-1981 (precatalytic converter) passenger vehicle produces three times the HCs, twice the NO_x, and eight times the CO of a new vehicle (Anderson, 1990). Pre-1981 passenger vehicles account for 71% of all vehicular HC emissions, but make up only 40% of the fleet and account for an even lower percentage of VMTs.[9]

Slower replacement of older vehicles by new cars also tends to limit emissions reductions because emissions increase with vehicle age. MOBILE4 predicts that after 50,000 miles, every 10,000 miles results in an increase of 0.15 gram per mile in tailpipe HC. CARB also uses mileage adjustment factors: a 0.09 gram per mile increase in NMHC per 10,000 miles.

The aging of the fleet may actually have been caused, to some extent, by emissions standards on new vehicles. Crandall, Gruenspecht, Keeler, and Lave (1986) found (using an earlier version of EPA's MOBILE4 model) that higher new vehicle prices as a result of the 1981 tightening of emissions standards slowed vehicle scrappage, aging the stock enough to increase VOCs and CO emissions over that of a scenario where standards were not tightened, and that this perverse effect lasted for 5 years. Overall, they found that in 1982 the aging of the vehicle stock from its 1967–78 average resulted in VOCs, CO, and NO_x emissions being 26%, 23%, and 11% larger, respectively, than they would have been had no aging occurred.[10]

Miles per gallon. Fuel economy of the average auto in the fleet has increased from 12.4 MPG in 1980 to 20 MPG in 1988 (Motor Vehicle Manufacturers Association, 1990) in response to federal corporate average fuel econ-

---

[9] Passenger vehicles 10 or more years older than the current model-year are driven only 66% of the miles a new car is driven (S. Davis, Shonka, Anderson-Batiste, and Hu, 1989, pp. 3–15).

[10] Reduced choices of vehicle models as a result of the CAFE standards were also identified as reducing turnover.

omy standards, but these increases have had uncertain effects on total emissions. First, if vehicles use less fuel, HC emissions should fall, because these pollutants are produced from fuel components (as opposed to NO_x, which is a by-product of combustion). However, this engineering relationship does not appear historically, because both emissions and fuel economy are regulated. Second, the increase in fuel economy makes the effective price of driving lower, which tends to increase vehicle miles traveled. This increase in miles driven can offset engineering-based reductions in emissions per mile driven.

Gallons per trip (congestion). Congestion is important because vehicle speed affects emissions per mile. Overall, according to EPA, HC and CO emissions per mile fall rapidly at higher velocities, but NO_x emissions rise at a slightly decreasing rate above 35–40 MPH. Idling obviously results in more emissions during a trip than if there were no idling, but emissions of NO_x at idle are negligible. CARB shows how a vehicle traveling 10 miles in 11 minutes will emit 2.5 times the running exhaust hydrocarbons[11] as one traveling the same distance in 30 minutes. Finally, accelerating and decelerating raise emissions of all types dramatically over that at a constant velocity. Thus, stop-and-go traffic produces much more pollution than cruising.

Note the importance of the differential effects of vehicle operation on HC and CO, on the one hand, and NO_x on the other. In cities where NO_x reductions are more productive at reducing ozone than VOC reductions, *diminishing* speeds (or, at least, not doing anything to increase speeds) may be more effective at reducing ozone than increasing speeds.

Considering many of these factors simultaneously, Cameron (1991), applying a transportation network and vehicle emissions model to Los Angeles (Harvey, 1983), estimates that increasing average speeds from 27 to 32 MPH (27%) reduces vehicular ROG and NO_x emissions 19% and carbon monoxide emissions 23% from baseline levels.

Because highways are getting more crowded, more of the time, HC emissions per mile are probably increasing, *ceteris paribus*. For major U.S. cities in 1988, Hanks and Lomax (1991) estimate that 37% and 46% of total peak period VMTs for freeways and arteries, respectively, occur during congested conditions, with nearly all major cities (Phoenix, Houston, and Detroit are exceptions) experiencing increases in these percentages since 1982. In California,

---

[11] Not counting cold start or hot soak emissions.

congestion is increasing at 15% per year. Average speeds on Los Angeles freeways are dropping dramatically, with the projection that by 2010, 50% of travel
will occur under congested conditions, as opposed to 10% today (Burmich,
1989). In Los Angeles' peak periods, Hanks and Lomax estimate that 75% of
VMTs on freeways and 50% of VMTs on principle arteries and streets took
place under congested conditions.

Trips per private vehicle. Because vehicle emissions vary significantly
with engine (and catalytic converter) temperature, the number of trips a vehicle
takes affects emissions. More specifically, the greater the number of cold starts
the higher the emissions, even if total VMTs are constant. For example, cold
start emissions produce 60% of total exhaust plus evaporative emissions (ignoring evaporative emissions while running) for a 5-mile trip. Even for a 10-
mile trip, cold start emissions are 50% of the total (Burmich, 1989).

Using National Personal Transportation Survey (NPTS) results from 1977
and 1990 (Liss, 1991), the number of trips has risen rapidly, although less rapidly than VMTs (46% versus 55% increase). This means that average miles per
trip has risen, from 8.3 in 1977 to 8.9 miles per trip in 1990. For emissions, the
implication of longer trips is that there are fewer cold starts per mile and, thus,
lower emissions per mile. However, the large increase in the number of trips
and miles traveled works in the other direction.[12]

To help fashion policy for reducing the number of vehicle trips, detail on
the composition of these trips is needed. From the 1990 NPTS, only 28% of
U.S. vehicle trips are work-related, a percentage that has been falling steadily.
Family and personal business generates 51% of trips, half of them for shopping, school/church and doctor/dentist, with the rest for "other personal business." In 1977, other personal business accounted for only 15% of total vehicle
trips, while in 1990, this category accounted for 24% of all trips. In contrast,
social and recreation trips have been falling slightly as a percentage of all trips.

The new NPTS has not yet been analyzed for time-of-day travel. Using the
1983 NPTS results, only 37% of trips occurred during peak periods (6 to
9 a.m., 4 to 7 p.m.), with 44% of trips from 9 a.m. to 4 p.m., the periods supporting over half of the family and personal business trips. In the South Coast,
40% of trips are at peak hours (peak is 6 to 9 a.m., 3 to 6 p.m.) (Burmich,

---

[12] Liss does not distinguish between a single trip and a series of trips chained together. Such a
distinction is needed to identify the number of "cold start" trips.

1989). Even during the weekday day, trips for family and personal business exceed those for work (43% to 39%) (Klinger and Kuzmyak, 1986, Table 5.7).

The 1990 NPTS figures also reveal that efforts at encouraging carpooling have been generally unsuccessful, with commuting occupancy rates actually falling slightly from 1.3 in 1977 to 1.1 in 1990. Occupancy per vehicle has also fallen more dramatically for family and personal business (from 2.0 in 1977 to 1.7 in 1990) and for social and recreational trips (from 2.4 to 1.8).

Number of vehicles and auto vehicle miles traveled. Finally, as a result of changes in the above factors plus an increase in vehicle ownership per household from 1.6 in 1977 to 1.8 in 1990, VMTs have increased dramatically, outdistancing the growth in licensed drivers. From 1977 to 1990, the latter increased by 28%, while VMTs increased 55%. Only a part of the increase in VMTs can be explained by an increase in VMTs per vehicle, which has risen only 13% over the decade (from 7,563 miles in 1977 to 8,539 miles in 1990) and was actually larger in 1970 (Motor Vehicle Manufacturers Association, 1990). Rather, much of the increase in total VMTs has come from use of additional vehicles. As new vehicle purchases per year are essentially flat, the increase in vehicles is a result of holding older vehicles longer. Brownstone and Lave (Chapter 2) attribute the increase in registered vehicles to an unparalleled increase in licensed drivers as a result of the baby-boomers reaching driving age and the rapid growth of women in the labor force. However, because these transitions have been played out, Brownstone and Lave find that "vehicle ownership is becoming saturated," and that future growth in VMTs will more closely track population growth. The implication for emissions is that an important reason for the lack of progress on vehicular emissions reductions may be eliminated.

Emissions by mode. Total emissions also depend on the mode used for trips (subscript k in identity [2]). We have focused exclusively, so far, on autos and light-duty trucks. Given that most trips (82%) are and have been in autos, with public transport playing an exceedingly small role (2.2%), this focus is warranted. However, if policies are to be considered that have the effect of shifting trips to other modes, the emissions characteristics of these modes need to be addressed. In this regard, at current levels of capacity utilization, buses, in comparison to autos, emit more diesel particulates and sulfur dioxide (SO_2) per passenger mile, but far less CO, VOCs, and somewhat less NO_x. Thus, if mode switching results in additional bus vehicle miles traveled (rather than simply raising occupancy rates on public transit), complex tradeoffs will present themselves as some pollutants decrease but others increase as a result of the switch.

Summary

To summarize this complex story, emissions have not fallen as far as they might have and ozone may not have improved because of the focus on regulating exhaust emissions rather than evaporative and running loss emissions, because of an increase in gasoline volatility, because of placing less attention on reducing NO_x emissions than VOC emissions, and because of a large increase in VMTs and a change in their character. People have more cars, drive them further each year, and use them on more trips (particularly nonwork and off-peak). These trips occur in progressively more congested conditions, with lower occupancy rates. Further, with new vehicle purchases flat, the increase in demand for vehicles has been met by older cars, that is, through slower retirement rates. In part, higher new car prices caused by pollution controls are to blame, although demographic and economic factors are surely driving the demand increase. The aging of the fleet has partly offset emissions reductions from replacing old with new vehicles; however, the completion of the demographic transition should slow vehicle demand, with effects on age of the stock unknown.

MOBILE SOURCE POLLUTION CONTROL POLICIES

Passage of the 1990 Clean Air Act Amendments has in some respects dramatically changed the regulatory landscape with respect to policies for the control of ozone-causing emissions from mobile sources: mandating development of gasoline reformulated to meet specific standards and its supply to the worst ozone nonattainment areas, mandating introduction of very clean vehicles into commercial fleets in some nonattainment areas, setting technological requirements for reducing refueling emissions and standards for evaporative emissions control, devising detailed provisions for conformity of transportation plans with the requirements of the CAAA, and requiring use of "enhanced" inspection and maintenance programs in the worst nonattainment areas. At the same time, the heart of the new amendments, the reliance on tighter emissions standards, has remained a focus throughout the Act's long history.

The new amendments also require areas violating the ambient air quality standards to make deep cuts in emissions (about 3% reduction in annual average VOC emissions per year). This provision will encourage the states to make an all out effort to identify strategies for reducing emissions. In addition, by actually requiring areas with the highest ambient ozone concentrations to adopt economic incentive strategies for reducing emissions if they cannot meet emission reduction milestones (and suggesting that areas with less extreme violations of the standard do the same), the Congress is signaling its interest in cost-effective approaches to bringing these emissions down. And, by requiring that local transportation plans be "in conformity" to the Amendments, it has put the transportation planning community on notice that the search for cost-effective emissions reduction solutions extends even to traditional planning of traffic flows, highway investments, mass transit systems, and the like.

The federal government and the states are looking at five broad types of control strategies to reduce vehicular emissions: emissions standards, development of new vehicle and fuel technologies, identifying and acting on high polluting vehicles, implementation of transportation controls (including changing total VMTs, their timing or location, and changing the modal split), and use of broader economic incentive measures (such as emissions or gasoline taxes). The advantages and disadvantages of these strategies are discussed below, with particular attention paid to their cost-effectiveness in reducing emissions of VOCs.

The cost-effectiveness of a strategy to reduce VOCs measures the cost of the strategy divided by the emissions reductions it brings relative to some base case. Thus, if strategy A has a *higher* "cost-effectiveness" estimate than strategy B, strategy A is *less* cost-effective than B. This simple measure breaks down when more than one pollutant is affected by a strategy. Strategies that are effective at reducing only one type of pollutant (e.g., a gasoline reformulated for lower VOCs) will be unambiguously less cost-effective than those that reduce all pollutants simultaneously (e.g., retirement of older vehicles), if the two strategies have equal control costs and the pollutant in common is being controlled to an equivalent degree.

However, in practice, these *ceteris paribus* conditions rarely hold. More common may be that strategy A delivers greater reductions of pollutant x but smaller reductions of pollutant y than strategy B. Absent any nonarbitrary means of allocating costs to each of the jointly controlled pollutants, the preferred approach is to monetize the effects of each and sum—that is, perform a cost-benefit analysis. Short of this approach, when considering only the ozone precursors (VOCs and NO_x), the joint cost allocation problem can be resolved by expressing "effectiveness" in terms of reductions in ambient ozone. However, this requires use of complex air quality models (as discussed above) and implies that cost-effectiveness would vary significantly across urban areas. Producing such estimates is beyond the scope of this paper.[13] In the cost-effectiveness literature, one finds either all costs assigned to the pollutant of interest or an *ad hoc* (e.g., 50/50) division of costs to the affected pollutants.

Emissions Standards

The primary strategy for reducing vehicular emissions has been that of setting tailpipe emissions standards on gasoline vehicles. This technology-forcing strategy has resulted in lower tailpipe emissions (as noted above) primarily through the technological advance of the catalytic converter. Indeed, some more recent advances have enabled some late-model-year vehicles to meet the 1994 standards set in the 1990 Clean Air Act Amendments (Sierra, 1989).

These new standards reduce exhaust emissions of HC [actually non-methane hydrocarbons (NMHC)] and NO_x, 39% and 69% of 1981 standards, respectively, beginning in 1994. An additional 50% reduction of the remaining exhaust emissions will probably be called for by 2004, depending on whether EPA determines that such reductions are needed to bring cities into compliance with the ambient standards.

The new amendments belatedly recognize the important role played by evaporative emissions by setting standards for their control and for the control of emissions from the refueling of vehicles. Overall, "improved" and "advanced" gasoline vehicles (running on conventional gasoline) are predicted to result in significant further reductions in HC emissions (U.S. Environmental Protection Agency, 1989; Sierra, 1989; Krupnick and Walls, 1992).

Whether these newest reductions in emissions standards will be cost-effective is an open question. One report (Environmental Reporter, 1991, p. 224) quotes the EPA as saying emissions controls to achieve the necessary HC

---

[13] See Krupnick and Walls (1992) for such a cost-effectiveness comparison between gasoline and methanol-fueled vehicles.

reduction of 0.216 gram per mile would cost $152, implying, for 100,000 miles of travel, cost-effectiveness of $6,400 per ton of HC removed.[14]

Whether this is too high depends on the cost-effectiveness of other options for emissions control and on the benefits of control. At least when compared to the cost-effectiveness of the previous tightening of HC standards in 1980 and 1981, this new round of tightening may be far less cost-effective. White (1982) estimated additional lifetime vehicle costs of $300 for 1980–1981 and more current vehicles relative to 1977–79 vehicles. Given emissions reductions of 0.1 ton HC per vehicle, cost-effectiveness is about $2,700 per ton of HC in 1981 dollars, or $3,500 per ton in 1988 dollars. EPA's most recent Cost of a Clean Environment report (Carlin, 1990) arrives at a similar estimate.

These cost-effectiveness estimates, as shown in Table 5-2, may be compared to a variety of other benchmarks. The Office of Technology Assessment (1989) estimated the cost-effectiveness of several HC control approaches. The cheapest approach was to reduce the volatility of gasoline, at $500 per ton, followed by controls on emissions from refueling, at $1,000 per ton. As additional reductions are needed, the Office of Technology Assessment (OTA) found that costs per ton rise steeply. Another useful benchmark is the benefits per ton of HC reduction. Krupnick (1988) and Krupnick and Portney (1991) estimated the acute health benefits of reducing ozone concentrations through reductions in VOCs in the northeastern U.S.[15] These benefits, which are based on sophisticated grid-based air quality modeling, EPA-approved dose-response functions, and survey estimates of the willingness-to-pay (WTP) to avoid such effects as "minor respiratory restricted activity days," cough days, and asthma attacks, range from $100 to $2,000 per ton VOC reduced, with a best estimate of about $500. By these benchmarks and subject to a host of caveats (e.g., the limited set of effects considered and the unsophisticated nature of the WTP surveys), the more recent tightening of the emissions standards has not been economic.

Indeed, this picture is even less attractive than it appears. Because emissions controls apply to all vehicles, including those sold and driven in "attainment" areas, much of the emissions reductions may bring little, if any,

---

[14] ($0.000152/mi.) / (0.216 g/mi. × 2.2 kg/lb. × 2,000 lb./ton) = $6,400 per ton.

[15] While there are other types of benefits associated with ozone reductions, these health effects are the most important because they underlie the NAAQS for ozone.

TABLE 5-2   Cost-effectiveness (CE) per Ton of HC Reduced from Passenger Vehicles.

Approach	Emission Reduction (g/mi.)	Cost/unit (current dollars)	CE (constant 1988 dollars)	Source
Emissions Standards				
1969 vs uncontrolled	7.0[a] (.7 mt)	30/vehicle	56	b
1980–81 vs 1977–79	1.0[a] (.10 mt)	300/vehicle	3,500	b
1994 vs 1981	0.216	152/vehicle	6,400[c]	Author
Gasoline Reformulations				
11 to 9 RVP	1.07	0.015/gal.	500	d
EC-1	0.94	0.02/gal.	4,300[e]	f
1992 Reformulated Gasoline	0.156	0.025/gal.	3,550[g]	h
Refueling				
Onboard	0.31	25/vehicle	1,100	d
Stage II	0.18	i	1,000	d
Alternative Vehicles				
M85 vs improved gas	0.472[j]	.475/gal.	33,000	k
M100 vs advanced gas	0.171	.309/gal.	60,000	k
Enhanced I&M Programs	l	m	2,200–8,300	d, n
			2,100–9,000	f, n
			7,000	o
Early Retirement	12.25	700/vehicle[p]	10,400	Author
TCM (Eliminate 10-mile trip[q])				
uncongested	17 grams	0.20 trip[r]	10,000[r]	Author
congested	26 grams	0.30 trip	10,000	Author
plus remaining cars	3.35 kg	0.36 trip	10,000	Author

[a] Change in exhaust and evaporative emissions. Reduction in metric tons in parentheses, assuming 100,000-mile vehicle life with discounting ignored [see White (1982) for an explanation].
[b] White (1982).
[c] Assuming 100,000-mile vehicle life.
[d] Office of Technology Assessment (1989).
[e] See text for explanation of calculation.
[f] Anderson (1990).

TABLE 5-2 (continued)
_____

g All VOC emissions included, from baseline of 1.04 g/mi. Estimates are preliminary.

h Environmental Protection Agency (1991c).

i Estimates only available on a per station basis.

j Per gram of ozone-forming potential ("reactivity"), a measure that permits formaldehyde and methanol emissions, as well as reactive hydrocarbon emissions, to be aggregated.

k Krupnick and Walls (1992).

l For OTA, 17% reduction in exhaust emissions. Complex emissions effects from other studies.

m For OTA, repair bills from $70–100 per violation; additional program costs of $14–35 per vehicle.

n Values rely, in part, on Sierra (1988).

o McConnell (1990).

p Assuming 5,000 miles left on vehicle.

q See text for explanation.

r A cost-effectiveness of $10,000 is assumed, with the implied cost of a trip (its reservation price) calculated by multiplying $10,000 by the estimated emissions reduction.

reduction in the already low ozone levels in these areas and, accordingly, confer few benefits to health.[16]

The emissions standards set under the CAAA apply to every state, except California, which, in recognition of unique atmospheric conditions in parts of the state, historically has been permitted to set tighter vehicle emissions standards than those set for the rest of the nation. One justification that has been made for uniform national standards lies in the notion of economies of scale— that production costs (and therefore vehicle prices) will be much higher because of reduced production of any given vehicle model if vehicle manufacturers had to meet different emissions standards in many different states.[17]

This argument rests on the virtually unexamined assumption that diseconomies to vehicle production would set in if more than two sets of emissions standards for gasoline vehicles were permitted. While letting each state adopt its own emissions standards might indeed result in major cost increases and create incentives for purchasing vehicles in more lenient jurisdictions, the point at which economies of scale are lost or are outweighed by the inefficiencies of jurisdiction shopping has not been identified. At the same time, there clearly could be benefits to giving local authorities more flexibility in cus-

_____

[16] Benefits to crops and forests may be registered in some of these rural and urban attainment areas, however.

[17] Another justification is that on equity grounds, no area of the country should be treated differently than any other.

tomizing their emissions reduction strategies (including setting emissions standards) to be consistent with local conditions.

Yet, there is some risk in permitting more flexibility. As can be seen in the response of a number of states who are following California's lead in adopting low-emission vehicle programs (see below), there may be a tendency to adopt the strictest standards, irrespective of their effectiveness and cost. It may not be efficient for New York and other states to be adopting California's tighter emissions standards when California's air quality problems are so much more severe.

Alternate Fuels and Vehicles

Attention has begun to turn away from reducing emissions of conventional gasoline vehicles to switching to new fuels and/or vehicle technologies. The 1990 CAAA require areas significantly violating federal ambient air quality standards to adopt low-emission vehicle programs, which in effect are programs to mandate introduction of alternate-fueled vehicles, and more far-reaching programs have been adopted by California and several states following California's lead. In addition, the CAAA specifies emissions and HC species composition requirements for reformulated gasoline applicable to some areas violating the standards and several states have "opted in" to the program even though their participation is not required. California has recently issued proposed specifications for its own reformulated gasoline program that would require such fuel to emit up to 70% less VOCs than the reformulated gasoline specified in the CAAA (New Fuels Report, 1991).

It is unclear whether this legislative and regulatory activity will have the desired effects of reducing violations of the air quality standards and raising the rate at which alternate fuels and vehicles become viable in the marketplace. Numerous reviews of the advantages and disadvantages of these vehicles and fuels (e.g., DeLuchi, Johnston, and Sperling, 1988) have so far failed to identify a clear frontrunner on grounds of costs and emissions reductions. One fuel-vehicle mix being particularly touted in the U.S.—methanol—has been estimated (Krupnick and Walls, 1992) to be a very costly option in terms of the reductions in ozone-forming potential that it delivers relative to "improved" and "advanced" gasoline vehicles—with a best estimate of costs for vehicles designed to run on 100% methanol of $60,000 per ton VOC emissions reduced.

The main drawback of relying on advances in vehicle technology for emissions reductions—and this applies to alternate-fuel vehicles as well as to gasoline vehicles—is that its effectiveness is dependent on turnover in the vehicle stock. This has two implications. First, this strategy is for the long-run, as such turnover takes time.[18] Second, if the new technologies are more expensive than the old and/or have characteristics that make the vehicles less attractive to consumers, the rate of vehicle turnover will slow (called new source bias), offsetting some of the emissions reductions obtained by the low emitting vehicles that are purchased. Losses in welfare could be particularly large if adoption of new vehicle and fuel technology is mandated (by banning or limiting sales of gasoline vehicles, for instance).

It is worth noting, however, that one "new" type of fuel—reformulated gasoline—has a key advantage over the other alternate vehicles and fuels: it can be used in older as well as new vehicles. Whether gasoline reformulations can be devised that are cost-effective is still uncertain, however. ARCO's EC-1 formulation for pre-catalytic vehicles has a cost-effectiveness in reducing hydrocarbons of about $4,300 per ton.[19] Formulations applicable to newer vehicles face an even tougher challenge, in that they must meet specific emissions and species composition requirements and wrench substantial emissions reductions out of relatively clean vehicles. The reformulated gasoline to meet the requirements of the 1990 CAAA has been estimated to result in a cost-effectiveness of about $3,550 per ton of VOCs reduced (1988 dollars) (U.S. Environmental Protection Agency, 1991c) and California's much cleaner proposed "phase II" reformulated gasoline has been estimated by the California Air Resources Board (New Fuels Report, 1991) to cost from $8,000 to $12,000 per ton of VOCs reduced. While incentives for technological advances could lower these costs, unanticipated increases in the demand for reformulated gasoline, such as could arise if additional states "opt in" to the reformulated gasoline program, could increase costs over projections.

Whether or not the alternate fuels and vehicles appear to be cost-effective given current and prospective technologies, proponents of low emissions

---

[18] Retrofit options typically have not been emphasized, although they could obviously affect the rate of implementation and the cost.

[19] Computed for a 1979-model-year vehicle with baseline tailpipe plus evaporative emission of 4.7 grams per mile HC (Anderson 1990) and a fuel efficiency of 20.25 MPG (Crandall, Gruenspecht, Keeler, and Lave, 1986). EC-1 is assumed to deliver a 20% reduction in these emissions for 2 cents per gallon (Anderson, 1990).

vehicle programs hope that by creating captive markets the private sector will be induced to step up its research, development, and demonstration programs, thereby producing products that can directly challenge gasoline vehicles in the marketplace. Public pronouncements about new breakthroughs in electric vehicles (EVs), for instance, suggest some optimism in this regard.

Of some concern, however, are attempts to create massive programs to subsidize the development of alternate-fueled vehicles, and, in particular, those that would subsidize one particular type of vehicle at the expense of others. A bill in Congress that would subsidize demonstration programs for electric vehicles by making up the difference between the EV retail price and that of a comparable gasoline vehicle (HR 1538) is a case in point. First, it is not clear that any subsidy of these vehicles is required for their commercialization. Life-cycle cost analyses of these vehicles are highly uncertain, but a recent analysis concluded that with current technology they could be cost-competitive with gasoline vehicles over the vehicles' lifetime, although the EV purchase price (including the battery) will be higher (DeLuchi, Wang, and Sperling, 1989; see Chapter 4).

Second, if a subsidy is to be given, it should not have the effect of guaranteeing this cost competitiveness in terms of retail prices of the vehicles. By offering this guarantee, no incentive is provided to the EV producer for designing and demonstrating a cost-competitive vehicle. Indeed, this type of program could skew research towards producing vehicles with attractive, but overly costly, performance characteristics. Once the subsidy is removed, such vehicles could be too expensive to be competitive.

If the vehicles are to be subsidized, a better idea is to make a lump sum payment to the producer for each vehicle sold. The amount of the payment could be set according to a study estimating the retail price or, even better, the expected lifecycle cost differential between an EV and a GV. Producers would be free to set any price they wanted on the vehicle. But under this plan, they have the incentive to make a cheaper vehicle, competing on price as well as vehicle characteristics. The more they sell, the more money they keep. If they can produce an EV with the same retail price as a GV, they keep the entire subsidy.

Targeting High Polluting Vehicles

Relatively little effort has been made to identify and mitigate emissions from "high-polluting" vehicles. Indeed, not until recently has it been recognized that very few vehicles contribute most of the emissions and that these vehicles may appear in all model-year classes. ARCO found that the most polluting of 16 pre-1981 vehicles produced nearly ten times the HCs of the least-polluting. GM has also found wide vehicle-to-vehicle variation in emissions, with the distribution of emissions roughly log-normal: 10% of 1986-model-year vehicles violate the HC standard at 50,000 miles, 27% violate the CO standard and 7% violate the NO_x standard (Haskew and Gumbleton, 1988). Stedman (1989) used remote sensing techniques to find that 10% of the vehicles in his Colorado sample were responsible for 50% of the CO emissions, and that these vehicles were generally older and not well maintained. Hansen and Rosen (1990) used remote sensing to find that 20% of autos measured in Berkeley, California were responsible for 65% of aerosol black carbon.[20, 21]

Most nonattainment areas have vehicle inspection and maintenance programs, which are required in cities violating the NAAQS. And the 1990 CAAA require that these programs be "enhanced" in some 74 areas violating the ambient ozone standard. At the least, the current programs have several drawbacks that limit their cost-effectiveness to an estimated $2,000 to $9,000 per ton of HC reduced (Anderson, 1990).[22] For one, they test vehicles while they are idle and warm rather than running or cold, when most of the emissions are emitted; two, some do not have strict enough penalties for failure to correct violations or even failure to have the vehicle inspected; three, pass rates are very high, meaning that administrative costs are large to catch the few violators; four, those that fail are not required to spend more than a set limit on repairs (generally, from $50 to $100). This "waiver" feature leaves vehicles with serious pollution problems still on the road. And fifth, very old vehicles are exempted from the programs.

---

[20] Sometimes called elemental carbon, this type of particulate contributes to visibility extinction.

[21] Note that these studies do not count idle and cold-start emissions, and may not fully count emissions from acceleration and deceleration.

[22] These estimates include both out-of-pocket and time costs. The upper end of this range seems more likely, according to Anderson (1990). McConnell (1991) places the likely cost-effectiveness even higher.

There are a variety of options for increasing the cost-effectiveness of these programs. These include introducing tampering checks into programs without them, raising waiver limits, tightening the emissions rates to pass inspection, and not exempting any vehicles. The CAAA require that enhanced inspection and maintenance programs address some of these: raising waiver limits to $450, inspecting for tampering, and using "on-road" emissions testing devices. The cost-effectiveness of such programs has not yet been thoroughly analyzed, primarily because the final rules have not been written. EPA has offered a pre-liminary estimate of only $500 per ton VOC reduction for an enhanced, bien-nial program using centralized facilities with dynamometers to simulate in-use emissions (Clean Air Report, 1991). But this estimate, at a minimum, ignores the increased time cost associated with an enhanced program (10–15 minutes more than current programs) plus additional inconvenience associated with centralized facilities in states with decentralized programs.

There are other ways of targeting high emitters. Raising vehicle registra-tion fees is a promising option because it should encourage scrapping older, lower valued vehicles. According to Anderson (1990), Japan charges from $250 to $500 per vehicle and has a much younger vehicle stock than the U.S. (8.4% of vehicles in Japan are 10 years old or more versus 30% in the U.S.), where fees range from $15 to $40.[23] A policy of differentiated fees, that is, higher fees for older vehicles, would be even better targeted for pollution reductions.

A possible drawback of any policies resulting in greater costs borne by owners of older or less well-maintained vehicles is that the policies may appear (and be) regressive. Lower income households own proportionally more older vehicles. From a 1985 survey (Anderson, 1990), 67% of the vehicles owned by households with annual incomes below $15,000 were more than 6 years old; for households making more than $15,000 annually, only 46% of their vehicles were more than 6 years old.

Programs to encourage retirement of older or dirty vehicles by offering bounties address the regressivity problem. Such vehicle retirement programs are being tested in California. The most interesting involves the purchase of pre-1971 vehicles by Unocal Corporation (1991) for $700 each, for what it hopes will count as emissions credit against emissions reductions it is required

---

[23] Faster economic growth in Japan doubtlessly accounts for much of the difference in fleet age. That Japanese auto insurance rates do not decline with age is another reason.

to provide from its industrial operations. While cost-effectiveness is high, at over $10,000 per ton,[24] this privatization of pollution reductions could deliver large cost-savings to the company, estimated to be about $161,000 per ton of VOCs (Air and Water Pollution Report, 1990). Given the emissions requirement faced by the company, society clearly benefits by permitting the company to obtain its emission reductions from early vehicle retirement, although at a price of $700 per vehicle, the money could perhaps be better spent obtaining emissions reductions elsewhere, say by paying for a portion of repairs to vehicles failing inspection and maintenance tests.[25]

This program has other potential drawbacks. If the vehicles purchased were to be scrapped soon anyway or, worse yet, were not being driven, then no emissions reductions would be purchased. Also, as currently structured the actual emissions of the vehicle are not taken into account. By basing vehicle purchase prices on emission test results, emissions reductions could be obtained more cost-effectively. Finally, it may be that older vehicles with emissions control technologies could be repaired at a cost far lower than the retirement price. Thus, preliminary screening could be called for. Anderson (1990) suggests that only vehicles failing the inspection and maintenance tests and that cannot be repaired within the waiver limits be eligible for the early retirement program. If repairs of $100 would reduce exhaust emissions by 80% of uncontrolled levels for 5,000 miles of driving, for instance, cost-effectiveness would be only $2,500 per ton.

In the longer term, it may be possible to identify the "high emitters" on the road, subjecting them to special inspections, fines, or other measures. Technologies have been field tested that measure emissions of vehicles (in one second) as they enter freeways and videotape the license plates of vehicles with high readings (Stedman, Bishop, and Peterson, no date). More far-reaching,

---

[24] Assumes a retired vehicle would have had uncontrolled HC exhaust emissions of 9 grams per mile (Shih, 1990) at 100,000 miles and evaporative emissions of 3.25 grams per mile (RVP = 11.0), and would have been driven 5,000 more miles before "natural retirement."

[25] The early retirement program may be significantly more cost-effective than these numbers would suggest. New information obtained from Unocal (1991) shows that for a sample of their scrapped vehicles, tailpipe emissions averaged over 16 grams per mile and evaporative emissions were estimated to be 8.5 grams per mile, twice the estimate for pre-1971 vehicle obtained from MOBILE4. Thus, cost-effectiveness of the retirement program is halved, at $5,200 per ton.

on-board computerized emission monitors could someday perform this task (among others).

Transportation Controls

The first part of this paper notes the importance of vehicle trips, particularly non-commuting trips, and speed (or congestion) in accounting for emissions from vehicles. Thus, any policy that reduces vehicle trips and/or congestion is potentially attractive for controlling HC emissions. Such policies are called transportation control measures (TCMs), examples of which are the traditional public spending-based policies of highway construction, building rail transit lines, and expanding and subsidizing bus services; and incentive-based policies to discourage driving, raise occupancy rates, and change the number, timing, and location of trips. Examples of the latter type of policy include raising downtown parking fees to discourage auto commutes, establishing high-occupancy vehicle (HOV) lanes to encourage carpooling, establishing alternate drive-days, and charging congestion fees. Obtaining cost-effectiveness es-timates for these policies is complicated because at least two types of effective-ness measures are relevant: reductions in travel times and reductions in emissions. Speed and emissions are nonlinearly related and costs are difficult to estimate since, in addition to out-of-pocket costs (e.g., added parking fees), effects on travel time and inconvenience need to be valued. Indeed, while TCMs generally have low out-of-pocket expenditures associated with them, nonmonetary costs, in terms of (say) inconvenience and longer commutes, may be inferred to be large, given the resistance to efforts at changing driving and overall commuting behavior.

A lack of information on willingness-to-pay to avoid inconvenience and maintain flexibility makes it difficult to estimate the cost-effectiveness of alter-native TCMs. The tack taken here is to ask what the willingness-to-pay to maintain one's current driving habits would have to be for the TCM to have a cost-effectiveness of no more than $10,000 per ton of VOC reduction. To give some sense of the interplay between emissions reductions and congestion, only one type of TCM is considered—eliminating trips through mandatory carpool-ing. A hypothetical case is developed and the willingness-to-pay associated with a cost-effectiveness of $10,000 is calculated by considering (1) only the direct emissions effects of eliminating a trip (both in uncongested and in con-gested conditions); (2) the emissions reductions that would arise for the vehicles remaining on the road that are able to operate in less congested conditions (i.e., at higher speeds); and (3) the time savings attributable to the improvements in congestion.

Turning to the first point above, eliminating a 10-mile trip for a 1987-model vehicle with 20,000 miles on it[26] reduces HC emissions by 17 grams (in the absence of congested conditions) (Burmich, 1989). To obtain a cost-effectiveness of no more than $10,000 per ton, the net cost of eliminating that trip (i.e., the willingness-to-pay to retain this trip—called the "trip reservation price" here) would need to be under 20 cents. Even if the trip took place under congested conditions (and, therefore, provided a reduction in HC of 26 rather than 17 grams), the trip reservation price could be no more than about 30 cents for the cost-effectiveness of trip elimination to be under $10,000 per ton HC reduced.

If enough trips were eliminated, congestion would fall for vehicles still making trips, which would lower their emissions (as well as reduce travel times). For some idea of the size of this emissions improvement and its effect on the trip reservation price for a $10,000 per ton HC reduction, the relationships between traffic volume, vehicle concentration, speed, and emissions must be considered. Papacostas (1987) provides a graph relating traffic volume (cars per lane per hour) to speed on a four-lane freeway with a design speed of 60 MPH (Figure 5-2).[27] At a volume of 1600 cars per lane per hour, average speed is 15 MPH. This translates to a vehicle concentration of 107 cars per mile.[28] Assume that a 5-mile section of freeway is experiencing these highly congested conditions during commuting hours. Eliminating 600 vehicles per hour (a volume of 1,000 vehicles per lane per hour) would permit speeds on this section to increase to 50 MPH, with a concentration of 20 vehicles per mile. Comparing these two cases, vehicles in the congested conditions take 20 minutes to travel the 5-mile stretch of road, while those traveling 50 MPH take six minutes.

Eliminating these 600 trips on this lane segment would reduce HC emissions from a 1981+ auto traveling on the segment by 0.71 gram per mile (Shih,

---

[26] "Eliminating" a trip could involve carpooling or chaining trips together rather than actually foregoing the trip altogether.

[27] This is an idealized situation that does not account for acceleration and deceleration.

[28] The fundamental equation linking these concepts is q=uk, where q is vehicle volume, u is speed, and k is concentration.

Average Speed (mph)

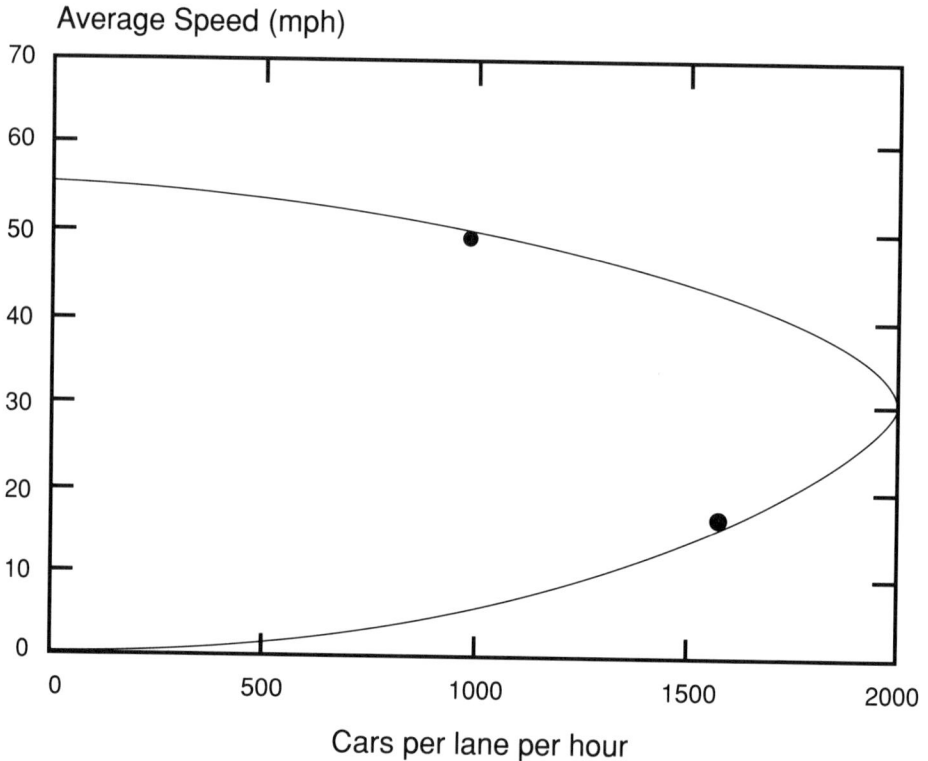

FIGURE 5-2   Speed-Volume Curve for 60 MPG Freeway. SOURCE: Papacostas (1987), p. 141.

1990) or 3.35 kg for the 1,000 vehicles remaining.[29] Converting into tons, dividing into $10,000 and dividing the result by 600 vehicles yields a figure of only 6 cents per vehicle trip eliminated. Comparing this figure to the 30 cents reservation price for the vehicle trip eliminated in congested conditions, it is clear that counting emissions reductions obtained through increased speeds of vehicles remaining on the freeway has little additional effect on cost-effectiveness. Overall, given the above examples, unless people would be willing to pay

---

[29] HC emissions at 15 MPH for a 1981+ vehicle at 50,000 miles are 0.72 g/mi. × 1.4 (speed adjustment factor) = 1.00 g/mi. At 50 MPH, HC emissions are 0.72 × 0.4 = 0.29 g/mi. The difference is 0.71 g/mi. Multiplying by 1,000 such vehicles and by 5 miles yields an emissions reduction of 3.35 kg HC per hour per lane. Note that these speed correction factors are averaged over a driving cycle, not specific to vehicle operating conditions.

less than 36 cents to retain a trip, trip elimination will not be less cost-effective than $10,000 per ton HC reduced.

Still, this example has ignored the time savings from vehicles remaining on the freeway segment. These savings raise the trip reservation price vastly beyond what it would be if only emissions reductions were counted. Saving the above 1,000 vehicles 14 minutes of driving time and valuing this time at $4.15 per occupant, results in benefits of $11.07 per vehicle trip eliminated, raising the trip reservation price with emissions reductions to $11.43.[30] Thus, in this example, the vast bulk of the benefits of TCM policies may be found in time savings; the emissions reductions that occur as joint products add little additional benefit. This implies that looking to emissions benefits as a major reason for supporting TCM could be looking in the wrong place.

Still, if TCM measures are designed that can efficiently reduce congestion, they will have some emissions reduction payoffs as well.[31] Of the myriad number of transportation control measures designed to reduce congestion,[32] some are particularly promising (and some are being considered in California). One approach is congestion tolls, or peak period pricing. This approach would charge vehicles by the location and/or time of their driving, providing monetary incentive to eliminate or shift trips (in time or space), to avoid congested highways. This approach has already been applied in Hong Kong (The Economist, 1989) where cars were fitted with license plates capable of being scanned by computers set up at key arteries within the city.[33] Cameron (1991) notes that the Coronado Bridge in San Diego replaced toll booths with scanners that can read special stickers or transponders in vehicles passing at high speeds. Vehicles using the bridge are assessed tolls via a monthly bill. Advances in such

---

[30] Winston (1985) reports auto commuting time valued at 178% of the wage. With an average hourly wage of $10, the 14-minute savings for the 600 vehicles carpooling with two people and the 400 vehicles with a single occupant, is valued at $6.64 per vehicle on average. Multiplying $6.64 by 1,000 vehicles and dividing by the 600 vehicle trips eliminated yields a value of time savings per trip eliminated of $11.07.

[31] Given the relatively deeper public support for measures to reduce emissions than for measures to reduce congestion, it may make sense politically to emphasize the emissions benefits of these measures.

[32] Under the CAAA, transportation policies in nonattainment areas will need to "conform" with reaching air quality goals. "Conformity" has yet to be defined, but will surely require, at a minimum, that emissions not increase as a result of new transportation strategies.

[33] The plan was ultimately rejected because of concern over what would happen to the collected revenues.

technologies and their merger with technologies being tried out in California to measure CO and HC emissions from vehicles waiting to enter Los Angeles freeways could eventually make it possible to charge vehicles for their contribution to congestion and identify high polluting vehicles at the same time.

Most efforts at implementing TCMs have focused on reducing commuting trips rather than daytime off-peak trips, which are a larger share of total trips. Reducing trips for shopping, transporting children to school and after-school activities, etc., may require different types of policies than those considered effective for reducing commuting trips. The centralization of both shopping and non-shopping activities in malls and the creation of new towns with such activities in walking or biking distance would have the benefit of encouraging the chaining of non-commuting trips. Of a more short-run nature, instituting parking fees with a high minimum charge at shopping centers might also encourage trip chaining, but would be unpopular with merchants and shoppers alike and, therefore, could not be implemented on a piecemeal basis.

In general, raising parking fees is a promising strategy. Cameron (1991) quotes a survey estimating that 90% of Southern California employers provide free parking for their employees. When some employers eliminated subsidized parking, ride-sharing increased between 18% and 83%. Cameron suggests that employers offer employees the cash-equivalent value of the parking space rather than eliminate parking subsidies outright.

While policies to encourage modal shifts away from passenger vehicles to diesel buses deliver joint benefits of reducing congestion and emissions of VOCs and CO, emissions of some other pollutants could increase (see above). Analysis of the relative costs and health benefits of reducing pollution from diesel versus gasoline engines is needed to identify these tradeoffs more effectively and to factor them into evaluations of mode-shifting policies.

Finally, as reducing health and other damages is the ultimate goal, policies that spatially or temporally rearrange emissions to reduce population exposures deserve attention. Thus, limiting traffic in the most built-up areas may increase traffic outside the area but population exposures to pollutants (and, therefore, health effects) may fall on net.

Emissions Fees, Gasoline Taxes, and Marketable Permits

The "mainline" economic incentive approaches—emissions fees and tradable emissions permits—are not well represented in the 1990 Clean Air Act Amendments for improving urban air quality through reducing vehicular emissions. There is a suggestion that these tools be used in ozone nonattainment areas that fail to meet emission reduction milestones.[34] And trading of emissions reductions among fuel suppliers in nonattainment areas is permitted.

California, however, is giving more attention to such approaches. As one example, CARB recently passed a vehicle credit trading program that establishes progressively tighter NMHC emission limits for a manufacturer's fleet, permitting manufacturers to earn credits for more than meeting the standards that can be sold to other manufacturers. Stevens (1991) analyzes this program and offers suggestions to improve its cost-effectiveness, while also presenting ideas for fuel credit trading and for emissions trading between point and mobile sources. California's proposed Drive+ program, if implemented, would tax new vehicles if their emissions exceed specified levels and provide rebates if emissions were below these levels (Guensler and Geraghty, 1991).

One reason for the general lack of attention to emissions fees on operating vehicles is that they may need to be quite large to influence driving and vehicle purchase behavior and that these costs will be born disproportionately by the poor, who tend to own the oldest (although not necessarily the dirtiest) vehicles. An important practical difficulty is in implementing emissions fees: not only would emissions need to be monitored, or a proxy for them devised, but the driver/owner of the vehicle would need to be periodically apprised of the fees he is incurring if driving behavior is to be modified. Ideally, an on-board device could be used to make such measurements and even report them to the driver. But the technology for such a device is still far off. Another approach would be to measure emissions on a dynamometer at the time of emissions inspection and bill on the basis of miles traveled since the previous inspection.

The monitoring difficulties inherent in this policy have led to suggestions for levying an additional gasoline tax. While such a tax has the advantage of being easy to implement, it too has some drawbacks as a proxy for an emissions tax. As pointed out by Davis, Grusky, and Sioshansi (1989), given that all vehicles must meet the identical emissions standards, a fuel efficient vehicle would pay a much lower tax than a gas guzzler while producing the same amount of emissions, making the implied tax per unit emissions much lower

---

[34] Areas defined as in "extreme" nonattainment, which only applies to Los Angeles at present, are *required* to implement economic incentive approaches.

for the fuel efficient vehicle. In addition, a gas tax is not sensitive to the timing or location of driving. Thus, it may not be very effective at reducing congested conditions.

In general, attempts to levy fees on any specific fuel, vehicle, or vehicle characteristic (such as fuel economy or carbon content) as a means of reducing emissions will introduce ancillary and possible perverse incentives. For instance, while the gasoline tax provides incentives to vehicle and fuel manufacturers to improve fuel economy, it could ultimately raise emissions. Increases in fuel economy lower the cost of travel, which leads to increases in VMTs, which, in turn, cause emissions to increase. Taxes on gasoline only, for instance, may reduce the attractiveness of gasoline vehicles but, by ignoring the polluting characteristics of all fuels, may not give enough incentive to develop and market the fuels that have the lowest social cost (private cost plus external cost).

CONCLUSIONS

There is both good news and bad in the above examination of policy approaches for reducing the contribution of vehicles to urban air pollution. The good news is that there are nearly limitless approaches and policy "levers" for influencing vehicular emissions and their spatial and temporal distribution. The bad news is that there is no single group of approaches that clearly stand out in front of the others in terms of cost-effectiveness (and non-regressivity), although reformulated gasoline, targeting high emitters, and policies influencing driving locations and timing (such as congestion tolls) appear promising. Identifying the most cost-effective approaches or the policies to permit their implementation will remain a daunting challenge.

The Clean Air Act Amendments of 1990 give little reason for hoping that this challenge will be met through federal efforts. To those who have seen the costs of mandated vehicular emissions reductions mount while ambient ozone levels in urban areas remain more or less unchanged, the wisdom of taking a further, major step down this path may be questioned. Subsidization or *de facto* mandating of alternative vehicle technologies seems a further step in the wrong direction. Leadership in applying economic incentives or other potentially more cost-effective approaches may come from California.

Finally, it may be that the NAAQS for ozone cannot be met at a cost that many would find reasonable. This possibility raises the issue of whether the NAAQS themselves are reasonable. Krupnick and Portney (1991) estimate that the acute health benefits to Los Angelenos or, alternatively, to the nation, of meeting the current NAAQS for ambient ozone are far outweighed by the costs, while acknowledging that significant uncertainties remain about the long-term effects of ozone exposure on the risk of developing chronic respiratory disease. Given that the NAAQS are set to protect the health of sensitive individuals (with a margin of safety), not to balance the health improvements in the general population against control costs, the existence of such inefficiency would not be surprising.

Whether efficiency considerations should dominate or even play a role in setting standards is not of concern here. There may be legitimate reasons for excluding the efficiency criterion in setting standards, having to do with the duty of government to protect public health, moral values, etc. However, such claims are far weaker when defending the arbitrary rule used to determine an area's attainment status—an area violates the standard when the fourth highest daily maximum monitored reading in three years exceeds it. If two days per year of violations were permitted instead of one, 22 of the 96 areas currently violating the ozone standard would be in compliance. And if six days of violations were acceptable, 69 areas would attain the standard. While acknowledging that relaxing this rule is a *de facto* relaxation of the standard, permitting efficiency considerations to have some impact on the definition of attainment would likely spare many areas the high costs of reducing ozone precursor emissions.

REFERENCES

Air and Water Pollution Report (1990). November 5. Silver Spring, Md.: Business Publishers, Inc.

Anderson, R. (1990). "Reducing Emissions from Older Vehicles." Report 053. August. Washington, D.C.: American Petroleum Institute.

Brownstone, D. and C. Lave (1992). "Transportation Energy Use." Chapter 2 of this volume.

Burmich, P. (1989). "The Air Pollution-Transportation Linkage." Report. Sacramento: California Air Resources Board, Office of Strategic Planning.

Cameron, M. (1991). "Transportation Efficiency: Tackling Southern California's Air Pollution and Congestion." March. Environmental Defense Fund and Regional Institute of Southern California.

Carlin, A. (1990). "Environmental Investments: The Cost of a Clean Environment. A Summary." Report EPA-230-12-90-084. December. Washington, D.C.: U.S. Environmental Protection Agency, Office of Policy, Planning and Evaluation.

Chang, T. Y., S. J. Rudy, G. Kuntasal, and R. A. Gorse, Jr. (1989). "Impact of Methanol Vehicles on Ozone Air Quality." *Atmospheric Environment*, Vol. 23, pp. 1629–1644.

Clean Air Report (1991). November 7. Washington, D.C.: Inside Washington Publishers.

Crandall, R. W., H. K. Gruenspecht, T. E. Keeler, and L. B. Lave (1986). *Regulating the Automobile*. Washington, D.C.: The Brookings Institution.

Davis, E. H., S. T. Grusky, and F. P. Sioshansi (1989). "Auto Emission Taxation: Alternative Policies for Improving Air Quality." Paper given at IAEE Conference in Los Angeles, October 16–18.

Davis, S. C., D. B. Shonka, G. J. Anderson-Batiste, and P. S. Hu (1989). "Transportation Energy Data Book: Edition 10." Oak Ridge National Laboratory Report ORNL-6565. September. Springfield, Va.: National Technical Information Service.

DeLuchi, M. A., R. A. Johnston, and D. Sperling (1988). "Methanol vs. Natural Gas Vehicles: A Comparison of Resource Supply, Performance, Emissions, Fuel Storage, Safety, Costs, and Transitions." SAE Report 881656. Prepared for International Fuels and Lubricants Meeting, Portland, Oregon, October 10–13. Warrendale, Pa.: Society of Automotive Engineers, Inc.

DeLuchi, M. A., Q. Wang, and D. Sperling (1989). "Electric Vehicles: Performance, Life-Cycle Costs, Emissions, and Recharging Requirements." *Transportation Research*, Vol. 23A, pp. 255–278.

Dowlatabadi, H., A. J. Krupnick, and A. Russell (1990). "Electric Vehicles and the Environment: Consequences for Emissions and Air Quality in Los Angeles and U.S. Regions." Discussion Paper QE91-01. Washington, D.C.: Resources for the Future, Quality of the Environment Division.

The Economist (1989). February 18.

Environmental Reporter (1991). May 24.

Guensler, R. and A. B. Geraghty (1991). "A Transportation/Air Quality Research Agenda for the 1990s." Paper 91-87.2, presented at the 84th meeting and exhibition, Air and Waste Management Association, Vancouver, British Columbia, June 16–21.

Hanks, J. W., Jr., and T. J. Lomax (1991). "Roadway Congestion in Major Urban Areas 1982 to 1988." Paper No. 910246, given at 70th Annual Meeting of the Transportation Research Board, Washington, D.C., January.

Hansen, A. D. A., and H. Rosen (1990). "Individual Measurements of the Emission Factor of Aerosol Black Carbon in Automobile Plumes." *Journal of the Air and Waste Management Association*, Vol. 40, No. 12, pp. 1654–1657.

Harvey, G. (1983). "Methodology for Incorporating Transportation System Effects Into Regional Transportation Energy Demand Forecasts." Report 400-82-023, prepared by Charles River Associates. Sacramento: California Energy Commission.

Haskew, H. M., and J. J. Gumbleton (1988). "GM's In-Use Emission Performance Past, Present, Future." SAE Technical Paper Series 881682. October. Warrendale, Pa.: Society of Automotive Engineers, Inc.

Klinger, D., and J. R. Kuzmyak (1986). "Personal Travel in the United States, Vol. I, 1983–1984 Nationwide Personal Transportation Study." August. Washington, D.C.: Department of Transportation, Federal Highway Administration.

Krupnick, A. J. (1988). "An Analysis of Selected Health Benefits from Reductions in Photochemical Oxidants in the Northeastern United States." In B. J. Morton, ed., final report prepared for the U.S. Environmental Protection Agency, Office of Air Quality Planning and Standards, Research Triangle Park, N.C. September. Washington, D.C.: Resources for the Future.

Krupnick, A. J., and P. R. Portney (1991). "Controlling Urban Air Pollution: A Benefit-Cost Assessment." *Science*, Vol. 252, pp. 522–528.

Krupnick, A. J. and M. A. Walls (1992). "The Cost-Effectiveness of Methanol for Reducing Motor Vehicle Emissions and Urban Ozone Levels." *Journal of Policy Analysis and Management* to be published.

Liss, S. (1991). "1990 NPTS: Nationwide Personal Transportation Study—Early Results." August. Washington, D.C.: U.S. Department of Transportation, Federal Highway Administration.

McConnell, V. D. (1990). "Costs and Benefits of Vehicle Inspection: a Case Study of the Maryland Region." *Journal of Environmental Management*, Vol. 30, pp. 1–15.

McConnell, V. D. (1991). "Assessment of the Cost-effectiveness of Vehicle Inspection and Maintenance Programs." Unpublished paper.

Milford, J. B., A. G. Russell, and G. J. McRae (1989). "A New Approach to Photochemical Pollution Control: Implications of Spatial Patterns in Pollutant Responses to Reductions in Nitrogen Oxides and Reactive Organic Gas Emissions." *Environmental Science and Technology*, Vol. 23, pp. 1290–1301.

Motor Vehicle Manufacturers Association of the United States, Inc. (1990). "MVMA Motor Vehicle Facts & Figures '90." Detroit: Motor Vehicle Manufacturers Association of the United States, Inc., Public Affairs Division.

National Research Council (1991). "Rethinking the Ozone Problem in Urban and Regional Air Pollution." Washington, D.C.: National Academy Press.

New Fuels Report (1991). October 7.

Office of Technology Assessment (1989). "Catching Our Breath: Next Steps for Reducing Urban Ozone." Report OTA-0-413. July. Washington, D.C.: Government Printing Office.

Papacostas, C. S. (1987). *Fundamentals of Transportation Engineering*. Englewood Cliffs, New Jersey: Prentice-Hall, Inc.

Pierson, W. R., A. W. Gertler, and R. L. Bradow (1990). "Comparison of the SCAQS Tunnel Study with Historical Data." Paper prepared for presentation at the 83rd Annual Meeting and Exhibition of the Air and Waste Management Association, Pittsburgh, Pa., June 24–29.

Russell, A. G., J. N. Harris, D. St. Pierre, and J. B. Milford (1989). "Quantitative Estimate of the Air Quality Impacts of Alternative Fuel Use." Final report to CARB. Sacramento: California Air Resources Board.

Shih, C. (1990). "MOBILE4 Sensitivity Analysis." Environmental Protection Agency report EPA-AA-TEB-90-02. May. Washington, D.C.: Government Printing Office.

Sierra Research, Inc. (1988). "The Feasibility and Costs of More Stringent Mobile Source Emission Controls." Report prepared for the U.S. Congress, Office of Technology Assessment. January 20. Sacramento.

Sierra Research, Inc. (1989). "Reactive Organic Gas Emission Factors and Fuel Economy Estimates for Advanced Technology Gasoline- and Methanol-Fueled Vehicles." Report No. SR89-10-01, prepared for Western States Petroleum Association. October. Sacramento.: Sierra Research, Inc.

Stedman, D. H. (1989). "Automobile carbon monoxide emission." *Environmental Science and Technology*, Vol. 23, No. 2, pp. 147–149.

Stedman, D. H., G. A. Bishop, and J. E. Peterson (No date). "On-road Remote Sensing and Motor Vehicle Inspection and Maintenance Programs." Unpublished paper. Denver: University of Denver, Chemistry Department.

Stevens, B. K. (1991). "A Tradeable Vehicle and Fuel Credits Program." Paper presented at the WEA International Conference, Seattle, Wash., June 29.

U.S. Environmental Protection Agency (1989). "Analysis of the Economic and Environmental Effects of Methanol as an Automotive Fuel." September. Special Report, Office of Mobile Sources. Washington, D.C.: Government Printing Office.

U.S. Environmental Protection Agency (1991a). "National Air Quality and Emissions Trends Report, 1989." Report EPA-450/4-91-003. February. Office of Air Quality Planning and Standards, Technical Support Division. Washington, D.C.: Government Printing Office.

U.S. Environmental Protection Agency (1991b). "National Air Pollutant Emission Estimates 1940–1989." Report EPA-450/4-91-004. March. Office of Air Quality Planning and Standards. Washington, D.C.: Government Printing Office.

U.S. Environmental Protection Agency (1991c). Note from Richard Rykowski to Kevin Neyland and Rich Theroux: "a Transmittal of Requested Economic Analyses Concerning Reformulated Gasoline" May 31. Washington, D.C.

Unocal Corporation (1991). "Scrap: A Clean Air Initiative from Unocal." Los Angeles.

White, L. J. (1982). *The Regulation of Air Pollutant Emissions from Motor Vehicles.* Washington, D.C.: American Enterprise Institute for Public Policy Research.

Williams, R. L., F. Lipari, and R. A. Potter (1989). "Formaldehyde, Methanol, and Hydrocarbon Emissions From Methanol-Fueled Cars." Report GMR-6728 ENV 271. June. General Motors Research Laboratories.

Winston, C. (1985). "Conceptual Developments in the Economics of Transportation: An Interpretive Study." *Journal of Economic Literature*, Vol. XXIII, pp. 57–94.

6 BALANCING ENERGY AND THE ENVIRONMENT

Margriet F. Caswell

INTRODUCTION

Background

The United States government, as the primary manager of natural resources, plays potentially conflicting roles in the development of domestic oil and gas reserves. The government is often the owner of energy resources and a partner in the profitable exploitation of those assets, but the government is also the custodian of environmental resources that may be harmed by energy development. During the past two decades, the American public has become aware of the importance of both resources. Energy resources are an important input to the U.S. economy, and there is an understanding that human welfare now and in the future may depend on the protection of fragile environmental ecosystems.

The battle lines between energy developers and environmentalists were drawn during the late 1960s. Since that time it apparently has been forgotten that a complex interaction exists among resources that precludes the "either/or" type of decision-making seemingly favored by extremists on both sides of the issue. Economic efficiency requires that resources be allocated so that the sum of joint profits be maximized. Such an allocation would maximize the net benefits to society. Both positive and negative interactions must be included within

Margriet F. Caswell is with Economic Research Service, Resources and Technology Division, Washington, D.C. The views expressed are those of the author and do not necessarily represent those of the Economic Research Service or the U.S. Department of Agriculture.

the decision-making process for each use of the environment—whether economic or aesthetic. Such a method effectively balances the costs and benefits for all of society. Tradeoffs and relative benefits should be explicitly included in any analysis. Physical, natural and social science aspects of each decision must be considered in order to determine the efficient management of every resource.

Oil and gas reserves are defined by laws, institutions, economics, and technology as well as geology. The oil included by the United States Geologic Survey (USGS) in the estimate of "reserves" is comprised only of those measured concentrations of petroleum that are feasible to extract under current economic conditions, given existing regulations and technologies. What is included within the inventory will change for many reasons even though the underlying geology would remain the same. For instance, exploration might confirm the existence of resources previously classed as "indicated," "inferred," or "hypothetical" by the USGS; legal or political restrictions to development might be altered; extraction technologies such as enhanced oil recovery methods might be improved; or petroleum prices might change. The decision to develop oil and gas reserves will also depend on each of these factors, which will ultimately determine profitability.

Blatant disregard for environmental resources (air, water, land, and biological resources) during petroleum development was never acceptable in America, although many abuses did occur. In the past, natural environments were deemed plentiful and the free disposal of wastes was considered to be good business practice. The ability of the environment to assimilate residuals has been surpassed in many areas, such that formerly accepted practices now impose costs on others. Economists refer to such cases as "externalities" and they represent a market failure. There is no market price for the use of the environment, so business may have no incentive to reduce waste disposal activities without some public intervention. Although there are sound economic reasons for government regulation to deal with externalities, there is the possibility that the direct and indirect costs of regulation will outweigh the benefits of reducing external effects.

To forbid energy development if there were any degradation of the environment, however, would be a devastating approach to solving environmental problems. The lack of petroleum resources would make the "luxury" of environmental protection and the maintenance of an acceptable life style hard to achieve. Choosing the optimal mix of regulations and incentives is difficult. Laws directly affecting oil and gas development will often have unintended impacts on the environment, and environmental regulations may either directly or indirectly affect the energy situation. Regulations and laws designed to protect one resource may inflict more costly damages on another. Direct and secondary impacts of government actions must be considered.

There will be tradeoffs involved in attaining a balance between petroleum development and a healthy environment. Unlike the choice of manufacturing two independent products, the energy/environment choice is complicated by the fact that oil and gas activities may have a negative external effect on the present and future opportunities to use the environment. The current federal regulatory structure evolved such that extractive and environmental resources are often managed by separate agencies, and laws are promulgated by different congressional committees. This structure led to a single-resource bias, which has resulted in excessive or contradictory regulations that may impose significant burdens on the petroleum industry and still not achieve any intended environmental goal. The stakes are very high in the perceived battle between oil and the environment, which makes it surprising that economic analyses have played such a small part in the decision making process during the last 20 years. The economically efficient development of natural resources—the allocation that maximizes net benefits to society—can be designed by using a systematic approach that includes all the primary and secondary effects of regulation on all parties. Regulators need to weigh their joint roles as developers and protectors of the nation's resources. Economic theory can be used to analyze the direct and indirect benefits and costs of proposed policies. In addition, the parties to which the benefits and costs accrue can be identified.

The potential physical effects of oil and gas development on air, water, and biological resources are briefly listed in the following section. Measurement of such effects in monetary terms is difficult, but the economic methodologies have been greatly improved.[1] Current policies concerning petroleum development have their genesis from two major events: the Santa Barbara oil spill of 1969 and the building of the Trans-Alaska Pipeline (1968-1977). These milestones are described below in the next section, and it is shown that perceptions may be more important than reality when decisions are made either in the

---

[1] Portney (1990) contains an excellent description of the ways by which the benefits and costs of environmental policies can be measured.

courts or in the political arena. The following section contains a description of two cases in which regulations promulgated by one government agency, ostensibly to protect the environment, may have increased the risk to other environmental resources. In the penultimate section of this chapter, a theoretical framework is described for the optimal accident prevention strategy, and then the reasons why such a policy might not be effective at this time are explored. The final section contains a plea for seeking a balanced policy based on the relative costs and benefits associated with both energy and environmental resources.

Possible Environmental Effects of Oil and Gas Development

The following list of potential external effects associated with petroleum development and use is presented to emphasize the myriad and complex interactions that are possible. The remainder of the chapter discusses only a subset of these possibilities to illustrate why balancing energy and the environment is a valuable endeavor. There are many negative external effects that the production, transport, refining, and consumption of petroleum products might have on the environment. An incomplete list is presented below merely to illustrate the potential impact of oil and gas development on air, water, and biological resources with concomitant effects on social welfare. Monetary values of impacts were not estimated for this chapter. Although the discussion is in terms of the receiving media, it must be remembered that environmental resources interact as well. For instance, a degradation of air or water quality can affect the health of biological resources.

Air Quality. Energy development and use can affect air quality in many ways. Both onshore and offshore production pollute the air with emissions from drilling equipment, escaping hydrocarbons, flaring of natural gas, and emissions from support vehicles. Pipelines contribute very little to air pollution during normal operations although there may be some emissions produced during transfer from wells or to refineries or tankers. Tankers and barge tugs emit hydrocarbons when under power and may have fugitive vapors escape during cargo transfer. However, the consumption of oil and gas contributes most greatly to air quality degradation. The incomplete combustion of petroleum products yields carbon monoxide, hydrocarbons, oxidation products, and particulates. The principal stationary-source contributors are "inorganic and organic chemical plants, iron and steel mills, petroleum refiners, pulp and paper mills, electric-power plants, and nonferrous-metal smelters. By weight alone, motor vehicles contribute nearly half of the total pollutants attributable to the

fact that carbon monoxide constitutes such a large proportion of emissions"
(Jones, 1978, p. 330). See chapters in this volume by Sperling (1992) and
Krupnick (1992) for discussions of automobile-induced air pollution and possible solution strategies. Air pollution can have local impacts (smog in Los Angeles), regional impacts (acid rain), or global impacts (greenhouse effects).
Hall (1990) calculates that the external cost of CO_2 generated from the burning
of oil to be $16.85 per barrel.

Water Quality. Degradation of water quality can occur in the marine environment or in surface or ground freshwater supplies. Both onshore and offshore production require the proper disposal of produced water.[2] Produced
water is the largest source of contaminants from offshore oil and gas development, where up to 1 barrel of produced water is obtained for each barrel of
crude oil that is recovered (Neff, Rabalais, and Boesch, 1987). Improper disposal can contaminate the marine environment or freshwater supplies. The disposal of drilling muds can also present problems, especially during offshore
production.[3]

Oil spills from offshore production and discharges from tankers and barges
have added many polluting constituents to water environments, and these
events are discussed at greater length below. From all sources, about 6 million
tons of oil pollutants get to the ocean each year (Mills and Toke, 1985). Consumption of petroleum products contributes to some water pollution indirectly
through the deposition of airborne contaminants. In addition, individuals dis-
posing of engine oil have unknowingly jeopardized their own water sources,
since one quart of crankcase oil that reaches a source of clean water will cause
250,000 gallons to be unusable for a drinking water supply (EarthWorks,
1989). Leaking gasoline storage tanks also can constitute a severe problem for
fresh groundwater supplies, and many service stations have been required to
replace tanks and remove contaminated soil.

Biological Resources. Both plants and animals can be affected directly by
petroleum development activities or indirectly through the air and water quality

---

[2] Produced water is the aqueous waste that has existed in the interstitial spaces of fossil fuel-
     bearing formations for thousands of years. Produced water may contain elevated concentra-
     tions of metals, hydrocarbons, and other organic substances.

[3] Drilling muds are used to remove cuttings from under the drill bit, to control well pressure, to
     cool and lubricate the drill string, and to seal the well. They can contain clay minerals, barite,
     trace metals, sodium hydroxide, biocides, diesel fuel, and other minor constituents.

changes discussed above. The integrity of an ecosystem can be destroyed if only one link in the food chain is eliminated. Therefore, environmentalists tend to focus on an indicator species such as the caribou in Alaska when protesting energy development projects. Such a focus can be misleading if the species chosen is not the most predictive indicator of ecosystem health. The species is often chosen because it is easily identified and tracked. Public opinion is often swayed by the careful choice of an indicator species—popularly known as a "charismatic macrofauna." Land-based production and transport can affect large territories during construction and by blocking migratory patterns with pipelines. Environmentalists also claim that oil production will open wilderness areas up for further development, thus endangering scarce biological resources.

Unfortunately, scientific knowledge of the complex interactions and dynamic nature of ecosystems is not adequate to accurately determine the effects of petroleum development on biological resources. Therefore, many environmentalists recommend erring on the side of caution (precluding development) until more is known. The relative benefits and costs of gaining more knowledge are seldom considered, however. Oil spills from onshore wells and pipelines have been relatively small and confined to a limited area. The greatest damage from such spills occurs when petroleum residues reach either surface or groundwater supplies. Offshore production spills and discharges from tankers and barges are potentially more damaging because they can contaminate large areas of the marine environment and directly threaten biological resources. In addition, little is known about the chronic exposure of marine resources to drilling muds and produced water that are the byproducts of offshore production. Neither refining nor consumption affect biological resources directly, although there is a possible secondary impact from degraded water and air.

Social and Economic Impacts. Environmental degradation has social and economic impacts with both monetary and nonmonetary components. Tourism and recreation industries and commercial fisheries will decline after an oil spill. Air pollution caused by the consumption of petroleum may deteriorate buildings. Health can be threatened by polluted air and water supplies and aesthetic values will be lessened when ecosystems are damaged. In addition, if the assimilative capacity of the environment to absorb wastes is decreased by oil development activities, other sectors of the economy may have to pay more to dispose of wastes.

The brief listing of possible environmental effects of oil and gas development was presented to indicate the complex interactions between human activities and natural resources. It is beyond the scope of this study to deal adequately with all aspects of petroleum development. Therefore, the following discussion will focus on the interaction between the production and transport of oil and the environment.

THE PERCEPTION OF OIL VERSUS THE ENVIRONMENT

There were two major events in U.S. oil and gas development history that defined the modern conflict between environmental and petroleum resources: the Santa Barbara Oil Spill and the building of the Trans-Alaska Pipeline (TAP). These events captured the attention of the American public and were used as examples of good (the environment) versus evil (oil companies). Although the actual events were not as damaging as initially feared, it was the public's perception of each case that colored the subsequent content of legislation and government action.

The Santa Barbara Oil Spill

On January 28, 1969, there was a blowout on platform A of tract 402 in the Dos Cuadras field in the Santa Barbara Channel off the Central California coast. Union Oil Company operated the rig for a group that included Texaco Incorporated, Gulf Oil Corporation, and Mobil Oil Corporation. The spill dumped 3.3 million gallons (approximately 79.3 thousand barrels) along 90 kilometers of coast (Allen, 1969).[4] Damages from the spill were estimated to be $16.4 million, which translates to a social cost of about $5 per gallon spilled (Mead and Sorenson, 1972).[5] When compared to the wreck of the *Torrey Canyon*, the

---

[4] The U.S. Geologic Survey's official estimate was that only 1.3 million gallons were spilled, but most sources, including the President's Panel on Oil Spills, use the Allen figure.

[5] The Mead and Sorenson study reported on the social costs of the spill for seven categories: (1) direct clean-up costs and property damage; (2) damage to tourism; (3) damage to the commercial fishing industry; (4) decline in real property values; (5) damage to the marine environment; (6) loss of the oil resource; and (7) aesthetic costs and reduction in recreational opportunities for the resident population. Conrad (1980) argued that $16.4 million was a conservative figure, but did not question the methodology that was used.

Santa Barbara spill was quite small. The *Torrey Canyon* was a tanker that went aground off the coast of Cornwall, England, on March 3, 1967, spilling 30 million gallons of oil on 140 miles of beaches in England and France. The fact that the Santa Barbara spill occurred soon after the *Torrey Canyon* incident was helpful for clean-up efforts and the mobilization of anti-oil organizations. Dispersants used on the European spill were found to be more harmful to the marine environment than the petroleum that was spilled, so such chemicals were not used on Santa Barbara beaches. Groups opposed to Outer Continental Shelf (OCS) development such as Get Oil Out (GOO) used strategies and data sources from newly formed European groups.

The Santa Barbara Channel is an area of natural petroleum seeps, so it has been argued that the marginal effect of the oil spill was quite small (Jones, 1969). The ecological effects of the spill never reached the extent that early observers had feared. Between 4,000 and 8,000 birds were killed and no wildlife populations were endangered (Straughan, 1972).[6] There is still no evidence that chronic long-term biological effects of exposure to the residual oil have damaged marine resources. Although potential long-term effects should not be ignored (Neushul, 1972), there is a danger that economic losses will be overstated if the ability of the ecosystem to recover naturally is not also included within the analysis (Anderson, 1983).

A 1990 Congressional Research Service report on the short- and long-term impacts of an oil spill concluded that species populations affected by the Santa Barbara blowout recovered swiftly and all environmental and socioeconomic consequences were relatively minor (Science, 1990). Such conclusions draw into question whether large expenditures should be made to save individual animals after an oil spill when the species population is not at risk. The money might be spent more effectively to address long-term environmental or social problems. Despite differing opinions of the severity of the event, Santa Barbara became the symbol for the environmental danger of OCS development.

Statements reflecting extreme views were the order of the day, and often received press coverage when "boring facts" did not. For instance, environmental historian, Roderick Nash, said,

> So I am prepared to say "no" to oil. Not "no, unless technology improves" or "no, until we really need the oil next century" or "no, if it involves more than

---

6   The reported number of birds killed has grown to over 10,000 in recent articles (Corwin 1990).

a dozen platforms," but just plain "NO!" . . . I reject the whole notion of compromise (Nash, 1970, p. 228).

Subsequent attempts to develop the petroleum resources found off the California coast have been strongly fought using the Santa Barbara spill as the rallying cry. It is clear that "Californians have separate and opposite sets of emotion for the oil that runs their cars and the oil that dirties what's left of their beaches" (Mead, 1978, p. 532). There have been no leases permitted in California state waters since 1969. All leasing activities have been outside the 1-mile limit since the Santa Barbara Spill.

The prevailing state and local consensus is that the probable environmental damages associated with OCS development outweigh the benefits. Federal leasing policies reflect the view that national benefits of development far outweigh national costs. Many of the benefits would accrue to those who would not bear the costs of an oil spill, however, so there is a distributional issue to consider when setting policy. Economic theory implies that if those that benefit from an action could compensate those that lose, then the action should be taken (a Pareto Improvement). Without actual compensation, however, a local community may be required to bear a large risk.

The Trans-Alaska Pipeline

Building the Trans-Alaska (TAP) pipeline was an incredible engineering feat. It also proved to be a legal, political, and economic feat as well. Although actual construction continued until 1977, most of the environmental battles were fought from 1968 until 1972.

On July 18, 1968, the Atlantic Richfield Oil Co. struck oil in Prudhoe Bay, Alaska, which is on the north slope of the Brooks Range at the edge of Arctic Ocean. There were indications that the Prudhoe Bay area contained between 10 and 12 billion barrels of recoverable high-grade oil. The amount of natural gas in and around the area was estimated at about 26 trillion cubic feet (Coombs, 1978). The oil discovery had an immediate impact on the state of Alaska. There had been three lease sales (1964, 1965, and 1967) that only netted lease bonuses to the state of about $12 million—an average of $12 per acre. The final lease sale took place on September 10, 1969, and netted the state $900 million at an average bonus price of $2,180 per acre (Roscow, 1977).

On February 10, 1969, Atlantic Richfield, British Petroleum, and Humble Oil announced that they would build the Alaska pipeline. Completion was expected in 1972, and the total cost of the system was estimated to be about $900 million (*Ibid.*). No time was wasted, and three days after the lease sale, the first freighter arrived at Valdez with pipe from Japanese steel mills.

The Trans-Alaska pipeline was planned to be nearly 800 miles long and have a capacity of carrying 2 million barrels (84 million gallons) of crude oil per day. The daily throughput rate was designed to be 1.5 million barrels supplied from 150 development wells at Prudhoe Bay. Over its entire route, pipeline construction would affect 30,000 acres out of the 586,412 square miles of Alaska. Environmentalists cautioned, however, that the pipeline would cross a fragile and delicate wildlife habitat that had scarcely been influenced by human activity.

The movement to protect the Alaskan environment came almost entirely from outside Alaska. Alaskans in general supported aggressive oil and gas development because of the revenues that would accrue. The State of Alaska had the incentive to speed oil development and to keep pipeline costs down, because the royalty payments to the state were based on the wellhead price of oil—the selling price at the refineries less all the costs of getting it there. The cheaper the pipeline, the greater the revenues for the state. The citizens of Alaska anticipated lower taxes and greater services from their state government share of oil revenue. This was a case of conflicting sovereignties that has continued. "Approximately 85% of the tax revenue raised by the State of Alaska comes from severance taxes and royalties on petroleum. Almost all of these taxes and royalties are paid by consumers and investors who live somewhere other than Alaska" (Jackstadt and Lee, 1990).

Saving the caribou became the *cause célèbre* of the development controversy. Environmentalists were not concerned solely for the caribou, but rather the ecosystem in which the caribou lived. Caribou were used as an indicator species and were thought to represent the wilderness. The biggest two groups of caribou in the pipeline area were the Arctic Herd, which numbered several hundred thousand and ranged west from Prudhoe Bay, and the Porcupine Herd, which was composed of around a hundred thousand animals and ranged east of Prudhoe Bay. The boundary of the herds was approximately the location of the pipeline in the north so the migratory patterns of these herds may not have been jeopardized in any appreciable way. Below the Yukon were two smaller herds —the Delta and the Nelchina—whose territories impinged on the pipeline route (Mead, 1978). The most critical period in the life-cycle of a caribou herd

is during the calving and post-calving aggregation period, which normally occurs between late May and mid-June.[7] Although little was initially known
about the caribou, strong opinions were expressed. Justice Berger of the
Supreme Court of British Columbia stated that

> "The preservation of the herd is incompatible with the building of a ... pipeline
> and the establishment of an energy corridor through its calving grounds. ...
> With the establishment of the corridor I foresee that, within our lifetime, this
> herd will be reduced to a remnant. ... Wilderness areas, if they are to be
> preserved, must be withdrawn from any form of industrial development. That
> principle must not be compromised" (Gray, 1979, pp. 136, 140).

The perceived risk of the pipeline was from the effect of that pipeline and
the presence of human beings on a large population of caribou. What was ignored, however, was the shooting of caribou that had been decimating their
numbers. For instance, the Forty Mile herd of caribou in Alaska, estimated at
500,000 in 1900, had been reduced by hunting to 8,000 by 1977. It was argued
that

> "for the state and federal officials to sanctimoniously enter into discourses as
> to whether a pregnant caribou will or will not cross a two-foot berm, while
> permitting almost indiscriminate shooting of these animals by any Alaskan
> along the migratory route, or any rich hunter wanting a 'double-shovel' set of
> antlers, is illogical" (*Ibid.*, p. 144).

The effects of hunting were largely ignored as environmentalists pushed to
prevent any change to the caribou populations caused by oil development no
matter what it cost.

Even without any demands from environmental interests, the pipeline con-
structors had a difficult engineering puzzle to solve. The stability of the ground
materials—whether soil or permafrost—in both arctic and subarctic regions
determined how the pipeline would be laid. There were three major methods:
ordinary burial (the cheapest), burial with insulation, and suspending above
ground. Heat transfer was the most important consideration. The oil comes out
of the ground at a temperature of about 160° Fahrenheit and loses only around
20° by the time it enters the northern end of the main pipeline (Coombs, 1978).
As it moves southward, friction heat that builds up as it is being pumped
through the long pipeline keeps it warm and free-flowing. The heat is a danger

---

[7] Caribou also travel to coastal areas, which are cool and windy during late summer, to escape
insect swarms that interfere with grazing.

to unstable permafrost, and large areas of tundra could turn into a quagmire if the permafrost melted. In these areas, ordinary burial could not be used. It was originally estimated that only 136 miles would require a raised pipeline (Roscow, 1977).

The original plans for the pipeline were thwarted by legislative changes. On January 1, 1970, President Nixon signed the National Environmental Policy Act (NEPA), which brought together under the Environmental Protection Agency (EPA) most federal environmental protection functions previously located in different departments. The EPA was not operational during most of the TAP negotiations, however. The Act also created the Council on Environmental Quality (CEQ) to serve as presidential advisors. NEPA required an Environmental Impact Statement (EIS) for major federal projects and laws (Section 102(2)c). The Act gave the government an unprecedented amount of control over projects. The pipeline project was not the catalyst for NEPA, but it was surely its proving ground.

The first EIS for TAP was submitted by the Department of Interior (DOI) in January of 1971. It contained 196 pages, plus 60 pages of construction stipulations. The total record of the required public hearings was over 12,000 pages long (Roscow, 1977). The EIS acknowledged that there were great environmental hazards of building a pipeline, but construction was approved anyway. As the result of a suit brought by the Wilderness Society, Friends of the Earth, and the Environmental Defense Fund, the court made the DOI do another EIS.

On March 20, 1972, DOI issued the second version of its EIS. This version now had nine thick volumes and dealt with the environment, design specifications, construction controls, and analyzed the economics of the project as well. On May 11, 1972 DOI Secretary Morton announced that he had decided to approve right-of-way permits for the Alaska pipeline. A pattern of environmental decision making was emerging in which "costs had nothing to do with the final decision. Public perceptions and political pressures had everything to do with the final outcome" (Gray, 1979, p. 3). The costs were felt to be borne only by the oil company.

The pipeline was planned so that it would stay clear of the migratory paths of the half million caribou that ranged throughout northern Alaska. Construction schedules were timed so as not to disturb their grazing habits or their calving grounds. In addition, the route was detoured around the nesting area of the peregrine falcon, an endangered species (Coombs, 1978). Conservationists demanded that above-ground pipe be elevated at least ten feet wherever animal migratory trails crossed the pipeline route. At that time, however, it still was not known what effect the pipeline would have on the caribou. Therefore, test sections were set up. It was reported that caribou, moose, and bison crossed under and even grazed beneath the elevated pipe without being frightened by its presence (*Ibid.*).

The pipeline designers preferred to protect the caribou by running the pipe underground, leaving a minimum opening of sixty feet through which the migrating animals could cross. Where the nature of the surrounding soil made this solution impractical, it would have been necessary to bury the pipe and refrigerate the ground around it to keep the soil from thawing—a costly arrangement. Because of permafrost and other ground conditions, 425 miles of the 800-mile pipeline are elevated (Roscow, 1977). By 1975, the construction committee re-estimated the project's still-rising costs at $6.4 billion. The cost was up almost $5 billion from the 1969 figure. Although costs were not broken down into categories, it has been estimated that more than $3 billion of the total was the result of inflation. Environmental expenses were estimated at up to $2 billion (*Ibid.*).[8] At that time, there were still two years of work to go until project completion. The final cost of the pipeline, terminal, and logistical support was about $8 billion (Mead, 1978).

As construction continued, more information was gathered about the Alaskan wildlife. Caribou were found to be "stolid" animals, phlegmatic creatures who are not easily spooked. Perceptions that caribou were very sensitive to the presence of humans may not have been accurate. A story is told about a group of caribou that walked through the middle of a camp site set up by a field party of biologists who were attempting to trace the migration patterns of the "illusive" caribou (Gray, 1979). Initial biological experiments had shown that the caribou (except for rare exceptions) would not cross under a pipeline, although moose readily would cross (Mead, 1978). After the pipeline's construction, however, it was found that caribou, instead of shying away from elevated pipe-

---

[8] It is unclear what costs should be assigned as environmental expenses. Suspending the above-ground sections of pipe at 10 feet rather than the planned 4 to 6 feet required costly engineering design modifications, but the estimated total cost of the 425 miles of above-ground pipe was only $1.1 billion (Mead 1978). Other environmental costs are less clear. The original estimates for the project may not have been realistic, which would also distort the calculations.

lines as expected, became accustomed to the pipe's presence and were often observed browsing beneath it (Coombs, 1978). Below the pipe the grass and tundra were often lusher than the surrounding area due to the proximity of the oil's warmth. It has been reported that the pipeline has increased forage areas and concentrated caribou and moose along the right-of-way (Hines, 1988). This may be a questionable benefit since the pipeline haul road would give easy access to "park and shoot" hunters.

The pipeline appears not to have caused serious disruption of caribou migrations (Carruthers, Jakimchuk, and Ferguson, 1984). Some changes in the caribou population were observed, however. Tens of thousands of caribou were counted in 1969 and 1970, but the 1971 and 1972 migrations seemed to drop to a tenth that many. In the following year the herds appeared in numbers approaching 20,000 (Mead, 1978). Biologists believe that heavy wind and snow conditions in the Brooks Range in the winters of 1971 and 1972 had determined how far north the caribou could range the next summers, not the presence of the oilmen (Roscow, 1977).

Angus Gavin, Arctic naturalist, zoologist and engineer wrote in 1972, "After five years of oil operation activity on the North Slope, it is quite evident that these operations have had little or no effect on the normal activities of the wildlife that frequent this part of the Slope" (*Ibid.*, p. 184). This outcome could be the result of very different forces. The caribou and the Arctic ecosystem as a whole may be much less fragile than environmentalists believed and/or the mitigation measures engineered into the pipeline design and regulated during construction may have been very effective. It is not possible to determine what the population of caribou would have been if the protection measures had not been undertaken. Whether the costs of these measures were justified by the benefits has not been addressed.

Although the preservation of caribou populations was a major focus of environmental concerns, the prevention of oil spills also received a great deal of attention. The pipeline builders had their own incentive to prevent spills, so they needed little urging from environmentalists to design for safety. Besides reinforcing every pipe joint, there are 62 massive gate valves installed along the line at strategic locations where a spill could cause significant damage to the environment. Most of the mainline gate valves can be remotely controlled from Valdez. In addition, there are 80 self-operating check valves that respond to pressure changes (Coombs, 1978). Alyeska Pipeline Service Company claims that from Prudhoe Bay to Valdez, there is not a mile of the pipeline route that hasn't been environmentally engineered in some way (Roscow, 1977).

The building of the TAP was a learning experience. A great deal of Arctic research was initiated that might not otherwise have been done for many years. Complex engineering problems were solved. It was understood, however, that the decision-making process concerning allocative choices between wilderness preservation and resource development was on trial. It was argued that "TAPS will have been more valuable even than the oil it carries if from its history we can derive more effectual principles of public policy and conduct and learn to apply them to the questions before us. We shall learn, or we shall perish; that is the natural order" (Mead, 1978, p. 543).

In 1977, the Department of Interior (which had the primary governmental responsibility for the TAP) held a conference for all interested parties to assess what had been learned about the decision-making process. The single-resource focus of agencies and laws was identified as a major contributor to the conflicting regulations that angered environmentalists, delayed development, and imposed costs on everyone. One conclusion reached at the conference was that all federal and state government bodies with a legal voice in the outcome be united in a single agency. It may not be realistic to expect perfect agency coordination, but improvements can be made. In addition, the conferees decided that the role of the judicial system in deciding essentially technical questions should be limited but not eliminated. "It is no more reasonable to allow the [court] final discretion in deciding, say, the engineering effects of permafrost than to suppose they can arrive at a medically valid diagnosis of small pox or cancer" (*Ibid.*, p. 561).

The public perception of the TAP is that the caribou have been saved (even though they may not have been in danger). The costs of complying with environmental regulations were not an important part of any discussion as long as the public was assured that the oil would eventually flow. There also seemed to have been an impression (at least up until the *Exxon Valdez* oil spill) that all environmental problems associated with Alaskan petroleum development had been solved.[9]

---

[9] Current discussions surrounding the proposal of opening portions of the Arctic Wilderness Reserve to oil development is reminiscent of arguments made over two decades ago—caribou herds must be allowed to freely migrate and wilderness access must be prevented at all costs.

There are many federal, state, and local environmental regulations in the United States that were enacted to reduce the negative external effects caused by oil and gas development, and many of these contain conflicting requirements. Compliance with these regulations involves costs for the petroleum industry. Even before the majority of environmental acts were in force, large sums were spent on pollution reduction. It was estimated that between 1965 and 1970 the petroleum industry spent an average of $1 million per day on all environmental mitigation measures, and in 1970 spent $288 million for the control or prevention of water pollution alone (Anderson, 1972). Sheppard (1982) estimated that the annual cost of complying in 1990 with stringent interpretations of the Clean Air Act, Outer Continental Shelf Lands Act, Coast Guard regulations, Deepwater Port Act, Safe Drinking Water Act, National Environmental Policy Act, Trans-Alaska Pipeline Authority Act, and several state air quality and production regulations would be $17 billion. The benefits of environmental improvement were not estimated in that study, nor were industry profits given, so there is no way to assess the relative impact of the environmental legislation.

Despite the existence of regulations, oil and gas development still is believed by many to be a threat to scarce environmental resources. The marine environment may be the environmental resource that is most endangered by oil and gas development and transport. Oil from spills or seeps approaches the coast by the mass transport of current and winds, and, in general, deposition is dominated by longshore currents rather than winds. Therefore, extensive lengths of coastline can be affected by relatively small quantities of oil. In most coastal waters the rate of spread of a near-shore oil spill is such that damage will occur before any action can be taken and it is unlikely whether recovery systems can capture more than 10% of the spill (Department of Environment, 1981; Science, 1990). Coastal areas are the habitats and breeding grounds for many biological species and these areas also provide recreation and aesthetic benefits for society. Therefore, a coastal spill can cause extensive and potentially costly damage.

The following section explores whether current resource allocation perceptions are formed or decisions are made any more rationally than they were before the enactment of the major pieces of environmental legislation. Two examples will be given of energy-environment conflicts in the coastal environment.

BALANCING OIL DEVELOPMENT AND ENVIRONMENTAL QUALITY

The focus of public attention on the coastal environment has often resulted in policies that shift external costs to another area or environmental media. The decision-making process currently in use still does not seem to rely on the balancing of benefits and costs. The reliance on perceptions to set the policy agenda may be a large barrier to sustaining environmental quality. Two examples are presented to show that the secondary effects of policies (which are often easily predictable) may have more significant effects than the initial intent. Both cases (the Long Beach to Midland Pipeline and the Point Arguello OCS development) deal with the choice of tanker versus pipeline, so a brief discussion of oil transport by tankers is presented.

Oil Transport by Tankers

In 1970, about 4,000 tankers moved over 600 million tons of crude oil each year (Anderson, 1972). Currently, most of the domestically produced crude oil carried by tanker originates in Alaska, and this oil represents the majority of coastal crude oil tankerage.[10] After the discovery of oil on the North Slope in 1968, there was a major debate concerning the best mode of transport for the oil from the Arctic area to the primary market in the Midwest of the United States. The Mineral Leasing Act of 1920 (Section 28) virtually banned the export of oil produced in the United States, so the domestic market was all that was open to producers of Alaskan oil. Two primary choices for transportation were discussed. The Trans-Canadian Pipeline (TCP) would have followed an inland route directly to refineries in the north-central United States. The Trans-Alaskan Pipeline (TAP) would end at the Port of Valdez in southern Alaska requiring tankers to transport the oil to refineries on the West Coast and the Gulf of Mexico. The greatest demand for the oil was in the Midwest, but after a long and heated debate, the TAP plus tanker alternative was chosen (Cicchetti, 1972).

The anti-export provision was also repeated within the Trans-Alaskan Pipeline Authorization Act (P.L. 93-153, 1973) thus requiring tankers to transport all of the Alaskan oil—about 1.7 million barrels per day (Mead, 1992)—south along the coast to the lower states for refining. Small amounts are

---

[10] Tankers carrying imported oil usually approach U.S. coastal waters only to unload cargo at designated ports. The volume of oil imports to the West Coast is very small (Mead 1991).

delivered to Washington and to the San Francisco Bay area, while the rest goes south to the Los Angeles Basin (a trip of about 2,000 miles along the environmentally sensitive coastline).

Refineries in Southern California initially could handle only about half of the Alaska production, so the remaining 800,000 barrels were transported to refineries in Texas (Hoyler, 1984). The crude oil was shipped in large tankers to the Gulf of Panama and then carried in smaller ships (lightering tankers) for the rest of the trip to Texas. The Prudhoe Bay oil field is being depleted rapidly, however, and production is estimated to fall by 50% during the 1990s (Mead, 1992). Therefore, the future amount of Alaskan oil that is shipped to Texas refineries will depend on further Alaskan oil development projects and Pacific OCS production.

Not only does the TAP Act require that Alaskan oil be shipped only to U.S. destinations, the Jones Act (1920) prevents the transport of goods between U.S. ports in anything but U.S. made, owned, and operated ships.[11] This law is an example of a regulation enacted for one purpose (i.e., to protect the domestic shipping industry) that has an effect on petroleum development and, as a consequence, also has an environmental impact. All Alaskan oil has to be shipped to U.S. ports on American flagged and manned ships because of the export ban and the Jones Act.

The economic efficiency of the export ban has been questioned. It has been suggested that an import-for-export plan with Japan would be profitable, and might reduce the risk to coastal environments (Cicchetti, 1972; Coombs, 1978). Alaskan oil is relatively heavy and the petroleum products demanded in California are best refined from lighter crude. Japan has a higher demand for heavier crude oil. If half of Alaskan output went to Japan and an equivalent amount of lighter crude was imported from Indonesia or directly from Nigeria to the Northeast, net imports would be the same, but profits would be higher. Mead (1992) reports on a study that estimated the removal of the export ban would increase wellhead price by $2.16 per barrel.

In addition to direct cost savings, the removal of the export ban would mean that tankers would spend more time on the open ocean rather than along the environmentally sensitive coastline. The strong sentiment against such trades was expressed by Alaska's Pipeline Coordinator who said, "If our oil

---

[11] Merchant Marine Act (41 Stat. 1007, 1920; U.S.C. 688, 1952).

goes to Japan, then foreign-bottom tankers can take it and therefore the safety and environmental standards will have been bypassed totally and we will have any leaky Greek that wants to come into our harbors to load oil" (Mead, 1978, p. 490). This statement reflects a common belief that foreign tankers would be inherently less safe than the more expensive U.S. ships. Alaskan State revenues would increase if transportation costs were lowered because the wellhead price for the oil would be higher, so those in other state agencies might disagree with the Pipeline Coordinator's statement.

Tankers contribute to water pollution in two ways: one, normal operating discharges, and two, accidental spills.[12] In 1983, tankers and barges transporting crude oil and petroleum products spilled almost two million gallons in United States waters while offshore operations only accounted for a spillage of 100,000 gallons and pipelines spilled less than 50,000 gallons. Long-haul and lightering tankers were responsible for only 150,000 gallons of the vessel spills while barges spilled close to 1.85 million gallons. In 1984, pipeline spills were negligible and offshore operations were responsible for less than 30 thousand gallons being spilled. Tankers spilled almost two million gallons and lightering tankers and barges accounted for an additional 2.5 million gallons (for a vessel total of 4.5 million gallons).

Water-borne transport as currently operated is much more likely to result in a spill than transport by pipeline. On a volume-per-mile basis, vessel transport of crude oil and petroleum products accounted for 3,000 gallons spilled per billion ton-miles in 1983 and almost 8,000 gallons in 1984. The amount spilled by pipelines was less than 100 gallons per billion ton-miles for both years (U.S. Coast Guard, 1987). Estimates of vessel movements have indicated that up to 80% of the accidents involve petroleum products, most of which are transported by inland tankers and barges (Department of the Environment, 1981). Current estimates of the amount of oil annually incorporated into the ocean by tankers during transport ranges from one to two million gallons (Shaw, Winslett, and Cross, 1987).

Transport is substantially more expensive when the oil is required to be carried on U.S. ships than if foreign vessels were used. United States shipbuilding costs are very high—at least twice as high as those in Japan. American crews command (through their strong labor union) high salaries, so the cost of

---

[12] Significant quantities of oil also can be released into the marine environment if fuel tanks are ruptured on any ship—tanker or nontanker.

labor is almost triple the cost for the most expensive international crew. Despite being more costly, the U.S. fleet does not have a particularly good safety record. A primary reason is that the U.S. tanker fleet is one of the oldest in the world.[13] From 1965 to 1975, while world tanker tonnage more than tripled, U.S. tonnage remained almost constant (Mead, 1978). Before TAP oil began to flow, the ships were relatively small because they were only needed for the coastal trade.

Oil tanker safety has been an environmental concern since the spills of the *Torrey Canyon* in 1968 and the *Amoco Cadiz* in 1978 (67 million gallons). The *Exxon Valdez* spill in American waters reinforced the perception held by the public that the environment would never be safe from oil pollution. Before the Santa Barbara spill in 1969, of the 32 spills ranging from 2,400 to 700,000 barrels, 21 were derived from tankers and barges. Tanker incidents accounted for 10 of the 13 spills greater than 30,000 barrels (Holmes and DeWitt, 1972). Ten years later, the general trend was the same. Tanker accidents accounted for about 135,000 tons of oil per year released into the ocean, while 790,000 tons were released through "normal operations" (Griffin and Steele, 1980). Current regulations prohibit all dumping of oil-contaminated ballast water into coastal waters. In the Alaskan operation, tankers keep the ballast water in their cargo compartments until they are berthed at Valdez. Then it is pumped to a treatment plant. The oil that is salvaged is added to the crude going into the tankers, and the purified water is returned to the sea (Coombs, 1978).

Despite evidence that up to 85% of the tanker accidents are caused by human error (Schuyler, 1982), regulations have usually been formulated to address technological factors. The current trend is to require double-hulled tankers. Economists have long argued that technology requirements are seldom efficient, and in this case, it is debatable whether double-hulling would increase safety (not to mention whether the environmental benefits of doing so would justify the cost of the requirement). In 1977, there was a serious proposal within the International Maritime Consultative Organization of the United Nations to make double bottoms and hulls *illegal* for crude-oil tankers because of an increased probability of sinking. In addition, double hulls would cost more and reduce the usable dimensions of the tankers (thus increasing the number of trips) (Mead, 1978). Analyses have shown that the number of trips or port-calls

---

[13] The fleet of Greece is older, but has only 6% of its tankers built before 1955 while the U.S. has 33% of its ships constructed prior to that time.

rather than the tonnage moved provides the best correlation with the incidence of spills (Beyer and Painter, 1977).[14] The potential size of a spill (given its occurrence) will depend on the carrying capacity of the vessel. Legislation that is written to specify design features for all new tankers above a certain size [e.g., 40,000 dead-weight tons (dwt)] or by a certain time encourages the shippers to use their smallest, oldest, and least expensive tankers.

The Jones Act has added substantially to the cost of shipping domestically produced oil, and the expected environmental benefits of using U.S. tankers may not exist. The safety of oil tanker transport is an important concern with respect to protecting marine and coastal resources.[15]

Long Beach to Midland Pipeline

Approximately half of Alaskan North Slope production initially was shipped to Texas because California refinery capacity was insufficient to handle the entire amount. The refined products were then shipped to the northern Midwest area of the U.S. where demand for heavy petroleum products was strongest. The Trans-Canadian Pipeline would have delivered the North Slope crude oil directly to the area of highest demand. The option was rejected on environmental and national security grounds. Shipping more than 1.5 million barrels of oil through West Coast waters and 800,000 barrels from California to refineries in Texas by tanker exposed coastal environments to a high level of risk. Pipelines have a better safety record than tankers with respect to spills. Losses from pipeline leakages amount to only about 0.006% of all oil transported (Griffin and Steele, 1980). Therefore, pipelines would be the preferred mode of petroleum transport if the goal was to minimize the environmental damages caused by oil spills.

In 1976, Sohio suggested using an abandoned natural gas pipeline to take the Alaskan North Slope oil from Southern California to Midland Texas. Sohio's preferred plan involved building a tanker terminal at Long Beach and deepening the port's ship channel to accommodate moderately big tankers (up to 100,000 dwt); buying the pipeline and the right-of-way; adding connecting

---

[14] If the number of trips is positively correlated with the use of smaller vessels, the type of ship will also be a factor in the number of spills.

[15] Studies of other industries such as timber production have shown that there would be substantial savings for suppliers and consumers if the Jones Act were repealed (Boyd 1983). The magnitude of those savings would depend on the availability of substitute products and the expansion characteristics of the market.

links; and supplying pumps.[16] The project itself was estimated to cost about $500 million and it would provide a capacity of 500,000 barrels per day.

Sohio secured the approval of dozens of federal, state, and local governments and agencies with a voice in the outcome, but the plan was blocked by the California Air Resources Board. The grounds for the refusal was that the unloading facility would discharge hydrocarbons and sulfur dioxide and thus degrade the quality of the atmosphere in the vicinity of Los Angeles.

Sohio agreed to obtain offsets for the additional pollution. An offset program would allow a new pollution source to operate in an area only if that source could reduce emissions at established sites enough to ensure that the resulting level of emissions would be *smaller* than before the new source was permitted. Such an offset policy results in a net improvement of air quality. Sohio promised to pay approximately $90 million for abatement at dry cleaning and generating plants.[17] It is believed that Sohio's plan would have "perceptibly improved the quality of Long Beach air," but the California Air Resources Board failed to approve the pipeline (Hines, 1988). Many regulatory agencies at that time were unfamiliar with offset programs and some environmentalists argued that no new pollution sources should be allowed under any circumstances.

Without the pipeline, the oil had to be shipped by tanker. The denial of the pipeline proposal is a case of one government body making a decision without reference to the others it affects. In order to supposedly protect the air resources of the Los Angeles area from the external effects of petroleum transport, the California Air Resources Board inflicted larger external effects on coastal resources in other geographical areas. Such parochial interests still hold sway in the decision-making process. Local interests may have believed that they must bear most of the risk and reap few of the benefits from the proposed pipeline. An integrated approach that considers all environmental resources and the resulting distribution of costs and benefits from a proposed resource use is seldom used.

---

[16] Gas pipe is generally thicker-walled than oil pipe, so the existing pipe was usable.

[17] Emission trading is an economically efficient method of reducing pollution since the reduction is done by the firms with the lowest marginal costs of abatement.

Point Arguello

The problems that arose in the development of offshore oil resources at Point Arguello in California also illustrate the conflicts that arise between decision-making agencies.

As onshore domestic petroleum resources become more difficult and expensive to exploit, the U.S. industry has turned to offshore fields to meet demands. Currently, the Outer Continental Shelf (OCS) is the source of 20% of total domestic oil and gas production. U.S. OCS resources are large, but unfortunately much of it is found in hostile climes. The Minerals Management Service estimates that undiscovered offshore oil resources in the Washington-Oregon planning area are only 36 million barrels, while the rest of the Pacific Region (California) is expected to yield 2.16 billion gallons (Cooke, 1984).[18] In comparison, the OCS area of Alaska may contain up to 15 billion barrels. The North Slope of Alaska had 12 billion barrels (Hines, 1988). OCS development has been the target of environmental protests throughout its history, particularly after the 1969 Santa Barbara oil spill. By that time, there were 8,000 offshore wells in federal waters and the safety records for those rigs were very good (Anderson, 1972). Between 1964 and 1973, there were 44 OCS development-related oil spills in Southern California that occurred in federal waters (beyond 3 miles offshore), which totalled 324,626 barrels. It was estimated that of about 2 million tons that had been spilled into the world's ocean by that time, less than 5% were contributed by offshore drilling and production operations, and more than 50% from tankers and ships (Holmes and DeWitt, 1972). The only other major OCS accidental spill was in 1978 at the Breton Island National Wildlife Reserve of Mississippi. Therefore, the reliance on tankers to import oil rather than developing local petroleum resources may present a higher probability for large accidental spills.

Communities adjacent to the OCS lease areas are concerned about the effects of petroleum production on their coastal resources. An offshore oil facility is in a fixed location and the local area will bear the external costs associated with a spill, whereas many of the benefits of OCS development occur else-

---

[18] The Minerals Management Service estimates are based on a large-scale computer program that generates point estimates. Solow and Broadhus (1989) have shown that these figures will overestimate the true value if underlying assumptions about the symmetry of the loss function are violated—a likely occurrence. Mead (1991) reports that California federal waters contain 1 billion barrels of oil reserves.

where.[19] In addition, there is currently insufficient scientific knowledge about the long-term chronic effects of drilling muds and produced water (byproducts of petroleum production) on the marine environment. The "Not in My Backyard" syndrome (NIMBY) is very strong and there is extreme political pressure to site OCS development "somewhere else." Bowing to such pressure in 1990, President Bush removed several large tracts from leasing activity until the year 2000 at the earliest. The development of as much as 2 billion barrels of oil and three trillion cubic feet of gas was thus postponed. The economic benefits to producers from such discoveries could be as high as $8 billion.[20] The one exception to the ban was the Santa Maria Basin area where development activities were already underway.

Point Arguello is located in the northeast portion of Santa Barbara County on the Central Coast of California just south of the Vandenberg Air Force Base. The OCS area is part of the Santa Maria Basin, which is rich in oil reserves. Oil was discovered in the area in 1981, and three oil platforms, an onshore cleaning plant, and a marine terminal comprise the Point Arguello project. These facilities have a production capacity of 75,000 barrels per day and were constructed at a cost of almost $2.5 billion. Original plans had proposed a larger capacity and the expected costs were considerably less. The need to comply with a continually changing set of environmental requirements significantly affected the development costs of the project.

There are several major pieces of legislation that directly affect OCS development and production and many others that indirectly impact the industry. The Outer Continental Shelf Lands Act (1978) illustrates the government's conflicting goals in resource management. The Act provides jurisdiction over submerged lands of the OCS and authorizes the Secretary of the Department of Interior to competitively lease such lands for resource production in order to capture a fair rate of return for the public asset. The Secretary must also determine rules and regulations deemed necessary and proper in order to prevent the waste of resources. The two goals—to maximize monetary returns and to minimize damages—may conflict. The OCS Lands Act places some of the burden of environmental protection directly on the petroleum industry by requiring

---

[19] The fixed location of the offshore oil facility also increases the effectiveness of accident response measures.

[20] Benefits will depend on future petroleum prices, production costs, and the costs of complying with regulations. The affected communities might not see any of the benefits directly.

that offshore oil and gas producers use the best available and safest technology (BAST). The Act also holds firms liable for damages from any offshore oil spill—thus lowering the expected monetary returns to the resource for both industry and the government.

In practice, however, the California Environmental Quality Act (CEQA) has had the greatest impact on the development of Point Arguello's resources. The relevant language of the Act (which is also included in the California Coastal Act of 1976) states that

> "each public agency shall mitigate or avoid the significant effects on the environment of projects it approves or carries out whenever it is feasible to do so. . . . In the event that economic, social, or other conditions make it infeasible to mitigate one or more significant effects of a project on the environment, such projects may nonetheless be approved or carried out at the discretion of the Public Agency, provided that the project is otherwise permissible under applicable laws and regulations" (Article 21002.2).

"Feasible" is defined as "capable of being accomplished in a successful manner within a reasonable period of time, taking into account economic, social, environmental, and technological factors." (Article 21061.1) The way that this provision has been interpreted is reminiscent of decisions made concerning the caribou in Alaska: All measures must be taken to prevent environmental degradation no matter what the cost as long as the oil will flow (i.e., there are positive profits). A company would be required to incur great costs no matter how small the benefits until all profits were dissipated. Only if the costs would result in losses (shutdown) would relative benefits and costs of mitigation measures be considered.

The completed Point Arguello project sat idle for several years due to local opposition to the oil being shipped by tanker to refineries in the Los Angeles area. The Santa Barbara County Board of Supervisors preferred that the crude oil be transported by pipeline for reasons of environmental safety. Chevron, which headed a consortium of 18 companies, argued that no suitable pipeline existed. The original plans for developing the Santa Maria Basin oil fields called for a pipeline that followed a coastal route directly to Los Angeles. Repeated attempts to obtain rights-of-way were thwarted by local communities, which wanted to avoid the "environmental disaster" of an oil spill. Information about pipeline safety and the probable lack of severity of a spill did nothing to dispel the impression that the construction of a pipeline represented an unacceptable risk for the communities involved. A decade of effort to have a pipeline corridor approved was a dismal failure. Each coastal community that denied permission did so ostensibly to protect their local environment despite the fact that the same oil would then likely be carried on tankers sailing directly off the community's coastline.

The only pipeline potentially available to carry the oil from the Point Arguello Project was the All American pipeline, which runs from the Gaviota marine terminal to Midland Texas. To get the oil to Los Angeles refineries would require diverting the oil south at Bakersfield, 100 miles inland. There are no heated pipelines between Bakersfield and Los Angeles, so the heavy crude oil extracted from the Santa Maria Basin would have to be mixed with lighter oil (possibly imported by tankers from Indonesia and then trucked from Los Angeles to Bakersfield) before being sent by pipeline to Los Angeles. Such a plan was estimated to cost $314 million over the life of the project (Corwin, 1990). These costs do not include external costs that would be borne by others such as those associated with the increase in air pollution caused by transporting oil by truck.

The Santa Barbara officials declared that the pipeline alternative was feasible. One supervisor stated, "Chevron's bottom line is to maximize profits. Our policy is to maximize the use of the pipeline. Chevron has a pipeline . . . but they tell us they don't want to use it" (Ibid.). Another supervisor said, "The transfer of wealth, the $314 million, . . . that's their [Chevron's] problem" (Lankford, 1990). At a public meeting, another official stated that the additional costs of the pipeline would only be considered of any consequence if the whole project was jeopardized. If the costs only meant less profits for the oil company, those costs were not included in the decision-making process. Such comments are in keeping with CEQA but reflect a misunderstanding of economic efficiency. Costs matter—if a company's resources are used unnecessarily (i.e., the costs of preventing pollution exceed the benefits of prevention) then those resources are unavailable for providing goods and services that are valued by society. As of this writing, the transport issue for Point Arguello oil has not been solved.

OPTIMAL POLICIES: INTERNALIZING POTENTIAL EXTERNALITIES

The environmental safety records for OCS oil production and the transport of crude oil by tanker and barge are different. Companies involved in OCS development invest heavily in accident prevention, which has reduced the expense of cleaning up spills. Economic incentives are believed to lead to efficient outcomes, and regulations can encourage or thwart these incentives. Firms respond to regulations that were designed to lower the probability of contamination (*ex ante* policy) and to the threat of suits for liability if a spill occurs (*ex post* policy). This section of the chapter will explore the economic and legislative factors that might explain why the coastal environment of the United States is more endangered by the ocean transport of oil than from offshore production.

Regulations can be of two types, command-and-control and performance-based. In the former, the regulatory agency determines the methods or technologies that the firm must use to prevent accidents. In the latter, the regulator sets a standard of safety that the firm must meet, but the firm may choose the methods that will be used to comply. In general, economists prefer performance-based standards because they encourage compliance at the least cost and offer incentives for technological improvement.[21]

Ex post liability also offers an incentive for pollution prevention. A firm can invest in accident prevention technology in order to reduce the probability of having to pay for damages incurred in the event of an accident. If the firm is required to pay the "external" costs of production, then the externality is "internalized." This does not imply that there will be no accidents or contamination, but the expected amount of damages would be acceptable (i.e., the costs of further prevention would outweigh the benefits). In common law, the term "strict liability" means that the firm must pay the penalty for polluting whether or not there was intent or fault involved. Under a negligence rule, the firm is found liable if they have not exercised a legal standard of care.

Several environmental studies have discussed *ex post* liability rules as a practical form of economic incentives for pollution damage control (Bradley, 1974; Conrad, 1980; England, 1988; Opaluch and Grigalunas, 1984). Shavell

---

[21] The costs of monitoring and enforcing regulations must be considered when choosing the most effective policy.

(1984), Segerson (1986), Johnson and Ulen (1986), and Kolstad, Ulen and Johnson (1990) have recommended a mix of *ex ante* regulations and *ex post* liability rules to produce the incentives necessary to reach the optimal amount of accident prevention. The following discussion of the effect of liability rules on OCS production and safety will summarize the arguments and conclusions presented in the articles mentioned above.[22]

In the simplest model, it is assumed that the oil producing firm facing *ex post* liability will choose the level of output and pollution control to maximize expected profits.[23] Pollution incidents will reduce profits at any time due to the payment for damages if an accident occurs and the reduction of the amount of oil to be sold. Accidents are assumed to be random events that are influenced by the actions of the firm.[24] In this example, there is no income constraint to prevent the full payment of damages by the firm.[25] The probability of an oil spill occurring is affected positively by the amount of output produced and negatively by the amount of accident prevention technology that is installed.

The expected profits for the firm are the net revenues earned from production less the costs of accident prevention and less the expected liability associated with accidents. The expected liability contains several key components. There is a known probability distribution of oil spills (number and size) that is affected by both oil production and the prevention technologies that are employed in a known way.[26] The monetary costs of each spill are made up of the value of the lost oil and the liability costs that would be assessed.[27] The firm has two decisions to make: Choose the amount of oil to produce; and choose

---

[22] The summary presented here does not include many of the important policy implications presented in the cited articles. The interested reader needs to obtain these sources to capture the richness of the analyses.

[23] Although the discussion refers to an oil-producing firm, the theory equally applies to a firm that transports oil. In the transport case, output would refer to the amount of oil that is shipped.

[24] For simplicity, no output price or cost uncertainty is included in the model.

[25] Shavell (1984) argues that liability will not create adequate incentive to take appropriate care if there is a possibility that the firm would not be able to pay for the harm done.

[26] The term "known" implies that all relevant parties have the same access to information, are in agreement as to the probabilities, and are all risk neutral.

[27] It is assumed that the liability costs would be equal to the true social damages caused by the accident. In practice, however, the determination of monetary damages is the most contentious part of the process.

the accident prevention technology. Profit maximization therefore requires that two conditions be met simultaneously. The first is that the level of oil production be chosen for a price-taking firm such that the benefits of producing another unit of oil (the marginal net revenue) be equal to the additional cost of producing that unit (including additional liability that would be expected by increasing production). The second condition is that the optimal amount of prevention technology will be chosen when the benefit to be gained from installing the additional unit (the reduction in expected liability) is equal to the cost of that unit. When these two conditions hold, both the level of oil production and the level of the externality are "optimal" from society's point of view.[28] Private profits for the company are reduced when the firm installs prevention technologies, but social benefits have increased when external effects are internalized. Consumer prices might increase due to these higher costs, but the price would then reflect the full social costs of oil production rather than just private costs.

There are obviously several problems with assuming that a model such as the one presented above can be used to represent an actual decision-making process. The probability distribution associated with the number and size of spills is not known (although we unfortunately are gaining more observations), and the effects of increasing output or adding safety technologies on the number and size of spills are not clearly understood. In addition, the environmental and social damages of oil spills are difficult to quantify. Kolstad, Ulen, and Johnson (1990) have shown that if the firm expects a liability cost less than the social optimum (due perhaps to a low number of successful prosecutions) then the firm will underinvest in accident prevention. Analogously, if the amount of liability were routinely set above the social optimum, then the amount of money spent on prevention would be too great. As discussed above, if costs outweigh the benefits (i.e., "too much" has been invested), then resources are unavailable for socially desirable investments.

Since uncertainty exists in so many areas of the firm's profit maximizing framework, it has been suggested that the government should be responsible for regulating accident prevention technologies. It can be shown that *ex ante* safety regulations will increase the level of investment in accident prevention. Shavell (1984) argues, however, that reliance on such *ex ante* regulations alone

---

[28] Using the Potential Pareto Improvement criterion, society is deemed better off if those that benefit from a change can compensate those that become worse off and still have a net gain.

will not result in the optimal level of prevention investment because the regulatory authority's information is imperfect.

One source of inadequate information is that the firm can take both observable and unobservable actions to reduce the risk of a spill (Segerson, 1986). The former might include installation of equipment or purchase of a double-hulled ship, and such actions would be relatively easy to enforce. The unobservable actions would include maintenance and operation procedures. As demonstrated by the *Exxon Valdez* accident, it is the actions that are unobservable *ex ante* that may be a large determinant of the probability of an accident. In order to be effective, such unobservable prevention measures must be enforced by the firm.[29] Therefore, a combination of *ex ante* technology regulation and *ex post* strict liability would be better than having either policy alone. The optimal mix will depend on the source and magnitude of uncertainty and on the relative costs of enforcement or litigation.

Both offshore oil development and oil tanker transport are subject to regulation and liability, but the effectiveness appears to differ. OCS facilities are owned by easily identifiable domestic firms, and the platform operations can be observed. In addition, the results of an accident can be readily attributed to the firm. The regulatory and legal system, although not perfect, seems to be working to reduce the risk of OCS spills.[30] Oil tankers, on the other hand, are more difficult to regulate, and damages caused by tanker accidents may be harder to assess. The U.S. can require that domestic or foreign ships entering federal waters meet certain technology requirements, but such regulations have high enforcement costs. The regulations are unenforceable outside coastal waters. With respect to the American tanker fleet (which carries all the Alaskan crude oil), conflicting government goals can be observed. The maintenance of a strong maritime industry is thought to be at odds with requiring expensive accident prevention technologies. High costs already have placed the U.S. fleet at a competitive disadvantage in the world tanker market. Higher costs would exacerbate the situation.

Oil-carrying tankers are subject to liability and must show proof of insurance. The liability applies only to the tanker owner, and not to the owner of the

---

[29] Segerson (1986) also points out that both industry and the public may be risk-averse, which implies that neither party should assume all the risk.

[30] The same characteristics hold true for onshore pipelines, which also have a good safety record.

cargo. Cleanup costs for oil spills are often very expensive and may greatly exceed the value of the vessel from which the oil was spilled. If the tanker represents all or most of the assets of the shipping company, there will be an income constraint with respect to the payment of the full liability, which will lower what the firm perceives the liability will be. The decreased expected liability cost would result in excess production (too much oil shipped) and too little prevention technology in use. The economic incentive that would prevail for an oil shipper would be to own a single, low-valued vessel rather than a fleet of well-maintained ships. If the liability extended to the owners of the oil, the income constraint would be less likely to obtain so that both production and safety precautions would be closer to the optimum. For the owner of the crude oil, the output decision process would remain the same—that is, to choose output such that the additional benefits of producing the last unit were equal to the cost. The choice of preventative technology, however, would included the choice of tanker. The tanker that would be chosen would be the one that fulfilled the second optimality condition above—that is, the reduction in expected liability based on the features of the ship (marginal benefit) would be equal to the additional cost. The existence of the Jones Act, however, would reduce the selection of tankers from which to choose.

CONCLUSIONS

This chapter was designed to illustrate the need for a balanced approach to the joint development of our environmental and energy resources. Both are valuable national assets, which are essential for maintaining a high quality of life. Since oil development, transport, and consumption activities may adversely affect the environment, there is a role for government involvement. Often, however, federal, state, and local agencies have had conflicting missions, so some laws designed to protect an environmental resource may have increased the risk of damage to other environments—leading to a net reduction in environmental quality. The direct costs of regulation and litigation may also dissipate any gains made by reducing externalities.

There is no way to eliminate the risk of oil contamination to the environment. Even if all production, transport, and use of petroleum was magically halted, natural oil seeps would continue to stain the oceans and beaches. The adversarial positions held by some energy development proponents and environmentalists must change. A reasonable balance will be reached when investments are made in accident prevention and pollution reduction such that the additional cost of making that investment equals the increased benefit from reducing the damage.

The model presented here suggests that a great number of uncertainties currently exist, which may prevent firms from internalizing the environmental damages that result from the production and transport of oil. Research will continue to reduce uncertainties concerning the effectiveness of accident prevention efforts and the physical damages caused by oil spills. Better methods for quantifying monetary damages need to be developed so that liability assessments reflect true social damages. As the uncertainties are reduced, oil producing and transport firms will be better able to choose spill prevention technologies that will approach the socially optimal level—and at the least cost. Although government regulation can then be reduced, there may always be a role for public oversight. The existence of unobservable prevention actions, risk aversion, and uncertainty lead to the need for a continued combination of technology regulation and liability rules. The challenge will be to develop a balance of policies based on the real costs and benefits associated with our energy and environmental resources.

REFERENCES

Allen, A. A. (1969). Hearings before the Subcommittee on Minerals, Materials and Fuels of the Committee on Interior and Insular Affairs, U.S. Senate, 91st, August. 1st Session, pp. 140–150.

Anderson, R. C. (1983). "Economic Perspectives on Oil Spill Damage Assessment." *Oil and Petrochemical Pollution.*

Anderson, R. O. (1972). "Environment and Social Goals." In R. W. Holmes and F. A. DeWitt Jr., eds., *Santa Barbara Oil Symposium—Offshore Petroleum Production: An Environmental Inquiry.* December 16–18, 1970. Washington, D.C.: Government Printing Office.

Barrows, T. D. (1972). "The Importance of Offshore Petroleum Resources." In R. W. Holmes and F. A. DeWitt Jr., eds., *Santa Barbara Oil Symposium—Offshore Petroleum Production: An Environmental Inquiry.* December 16–18, 1970. Washington, D.C.: Government Printing Office.

Beyer, A. K. and L. J. Painter (1977). "Estimating the Potential for Future Oil Spills from Tankers, Offshore Development, and On-Shore Pipelines." *Oil Spill Conference Proceedings*, pp. 21–30.

Boyd, R. (1983). "Lumber Transport and the Jones Act: A Multicommodity Spatial Equilibrium Analysis." *The Bell Journal of Economics*, Vol. 14, No. 1, pp. 202–212.

Bradley, P. G. (1974). "Marine Oil Spills: A Problem in Environmental Management." *Natural Resources Journal*, Vol. 14, pp. 337–359.

Carruthers, D. R., R. D. Jakimchuk, and S. H. Ferguson (1984). "The Relationship Between the Central Arctic Caribou and the Trans-Alaska Pipeline." Report to Alyeska Pipeline Service Company. April.

Cicchetti, C. (1972). *Alaskan Oil: Alternative Routes and Markets*, Baltimore, Md.: Johns Hopkins University Press.

Conrad, J. M. (1980). "Oil Spills: Policies for Prevention, Recovery, Compensation." *Public Policy*, Vol. 28, No. 2, pp. 143–170.

Cooke, L. W. (1985). "Estimates of Undiscovered Economically Recoverable Oil and Gas Resources for the Outer Continental Shelf as of July, 1984." OCS report MMS 85-0012. Washington, D.C.: U.S. Department of the Interior.

Coombs, C. (1978). *Pipeline Across Alaska*, New York: William Morrow and Co.

Corwin, M. (1990). "New Chevron Plan for Point Arguello Oil is Turned Down." *Los Angeles Times*, November 13.

Department of the Environment (Britain) (1981). "Oil Pollution of the Coastline."

The EarthWorks Group (1989). *50 Simple Things You Can Do to Save the Earth.* Berkeley, Calif.: Earthworks Press.

England, R. W. (1988). "Disaster-Prone Technologies, Environmental Risks, and Profit Maximization." *KYKLOS*, Vol. 41, pp. 379-395.

Gray, E. (1979). *Super Pipe: The Arctic Pipeline—World's Greatest Fiasco.* Toronto: Griffin House.

Griffin, J. M. and H. B. Steele (1980). *Energy Economics and Policy.* New York: Academic Press.

Hall, D.C. (1990). "Preliminary Estimates of Cumulative Private and External Costs of Energy." *Contemporary Policy Issues*, Vol. 8, No. 3, pp. 283-307.

Hines, L. G. (1988). *The Market, Energy, and the Environment.* Boston: Allyn and Bacon.

Holmes, R. W. and F. A. DeWitt, Jr., eds. (1972). *Santa Barbara Oil Symposium—Offshore Petroleum Production: An Environmental Inquiry.* December 16-18, 1970. Washington, D.C.: Government Printing Office.

Hoyler, M. (1984). "The Politics and Economics of Alaskan Oil Exports." In C. K. Ebinger and R. A. Morse, eds., *U.S.-Japanese Energy Relations: Cooperation and Competition.* Boulder, Colo.: Westview Press.

Jackstadt, S. and D. R. Lee (1990). "Tax Exporting and Run Away Government: The Case of Alaska." Working paper.

Johnson, G. V. and T. S. Ulen (1986). "Designing Public Policy towards Hazardous Wastes: The Role of Administrative Regulations and Legal Liability Rules." *American Journal of Agricultural Economics*, Vol. 68, No. 5, pp. 1266-1271.

Jones, C. O. (1978). *Clean Air: The Policies and Politics of Pollution Control.* Pittsburgh, Pa.: University of Pittsburgh Press.

Kolstad, C. D., T.S. Ulen, and G. V. Johnson (1990) "*Ex Post* Liability for Harm vs. *Ex Ante* Safety Regulation: Substitutes or Complements?" *American Economic Review*, Vol. 80, No. 4, pp. 888-901.

Krupnick, A. J. (1992). "Vehicle Emissions, Urban Smog, and Clean Air Policy." Chapter 5 of this volume.

Lankford, J. (1990). "Board Turns Down Tanker Plan." *Santa Barbara News Press*, November 13.

Mead, R. D. (1978). *Journeys Down the Line: Building the Trans-Alaska Pipeline.* Garden City, New Jersey: Doubleday.

Mead, W. J. (1992). "Crude Oil Supply and Demand." Chapter 3 of this volume.

Mead, W. J. and P. E. Sorenson (1972). "The Economic Cost of the Santa Barbara Oil Spill." In R. W. Holmes and F. A. DeWitt Jr., eds., *Santa Barbara Oil Symposium—Offshore Petroleum Production: An Environmental Inquiry*. December 16–18, 1970. Washington, D.C.: Government Printing Office.

Mills, R. and A. N. Toke (1985). *Energy, Economics, and the Environment*. Englewood Cliffs, New Jersey: Prentice-Hall.

Nash, R. (1972). "Progress and Poverty: The Santa Barbara Story." In R. W. Holmes and F. A. DeWitt Jr., eds., *Santa Barbara Oil Symposium—Offshore Petroleum Production: An Environmental Inquiry*. December 16–18, 1970. Washington, D.C.: Government Printing Office.

Neff, J. M., N. N. Rabalais, and D. F. Boesch (1987). "Offshore Oil and Gas Development Activities Potentially Causing Long-term Environmental Effects. In D. F. Boesch and N. N. Rabalais, eds., *Long-term Environmental Effects of Offshore Oil and Gas Development*. New York: Elsevier.

Neushul, M. (1970). "The Effects of Pollution on Populations of Intertidal and Subtidal Organisms in Southern California." In R. W. Holmes and F. A. DeWitt Jr., eds., *Santa Barbara Oil Symposium—Offshore Petroleum Production: An Environmental Inquiry*. December.

Opaluch, J. and T. Grigalunas (1984). "Controlling Stochastic Pollution Events Through Liability Rules: Some Evidence from OCS Leasing." *The Rand Journal of Economics*, Vol. 15, pp. 142–151.

Portney, P. R., ed. (1990). *Public Policies for Environmental Protection*. Washington, D.C.: Resources for the Future.

Roscow, J. P. (1977). *800 Miles to Valdez: The Building of the Alaska Pipeline*. Englewood Cliffs, New Jersey: Prentice-Hall.

Schuyler, A. H. (1982). "The Benefits and Costs of a Vessel Traffic System for the Santa Barbara Channel." Presented at the California Coastal Commission Symposium on Marine Resources Management, Asilomar, California, November 7–9.

Science (1990). "Spilled Oil Looks Worse on TV." Vol. 250, p. 371.

Segerson, K. (1986). "Risk Sharing in the Design of Environmental Policy." *American Journal of Agricultural Economics*, Vol. 68, pp. 1261–1265.

Shavell, S. (1984). "A Model of the Optimal Use of Liability and Safety Regulation." *Rand Journal of Economics*, Vol. 15, pp. 271–280.

Shaw, B., B. J. Winslett, and F. B. Cross (1987) "The Global Environment: A Proposal to Eliminate Marine Oil Pollution." *Natural Resources Journal*, Vol. 27, No. 1, pp. 157–185.

Sheppard, W. J. (1982). "The Cost of Environmental Regulations to the United States Petroleum Industry." *Oil and Petroleum Pollution*, Vol. 1, No. 1, pp. 49–57.

Solow, A. R., and J. M. Broadhus (1989). "Loss Functions in Estimating Offshore Oil Resources." *Resources and Energy*, Vol. 11, No. 1, pp. 81–92.

Sperling, D. (1992). "Alternative Fuels." Chapter 4 of this volume.

Straughan, D. (1972). "Ecological Effects of the Santa Barbara Oil Spill." In R. W. Holmes and F. A. DeWitt Jr., eds., *Santa Barbara Oil Symposium—Offshore Petroleum Production: An Environmental Inquiry*. December 16–18, 1970. Washington, D.C.: Government Printing Office.

U.S. Coast Guard (1987). Data Series.

7 OIL AND GAS LEASING POLICY ALTERNATIVES

Walter Mead

INTRODUCTION

The Iraqi invasion of Kuwait on August 2, 1990 and the subsequent loss of oil supplies from these two sources (5.2 million barrels per day or 9% of the world oil supply) has led to new concerns for oil supply reliability and renewals for a comprehensive national energy policy. For the United States, domestic crude production supplied 45% of the nation's crude requirements in 1990, with 55% of those requirements coming from imports (U.S. Department of Energy, 1991a). Domestic sources allow the U.S. to be relatively secure in its crude oil supplies compared to other leading industrial nations excepting portions of the former USSR and the United Kingdom. However, the U.S. Department of Energy forecasts that the U.S. will become increasingly import-dependent in the future with imports accounting for 62% of our requirements in the year 2000, increasing further to 68% by 2010. This dependency increases in spite of virtually no increase in consumption over the next 20 years as real prices increase from an average of $18.81 per barrel in 1990, to $34.20 per barrel in 2010 (1990 constant dollars) (U.S. Department of Energy, 1991b, p. 52).

The national security significance of these facts and forecasts is moderated by two conditions. First, while the U.S. is currently 45% import-dependent, only 30% of these imports are from OPEC, and a lesser share of imports is from the volatile OAPEC (Arab members of OPEC). These shares will grow in the future because crude reserves there are large relative to the United States and other non-OPEC countries. Second, as of February 1991, the U.S. Strategic

215

Petroleum Reserve (SPR) held 582 million barrels of crude oil. Even in the event of an unlikely 20% reduction in U.S. imports, SPR supplies would be sufficient to maintain normal refining levels for about 500 days.

The West Coast is currently in a secure supply position, with production available from Alaska and California. However, this security is rapidly diminishing due to declining Alaskan production, falling production of conventional crude oil from California onshore fields, plus continuing environmental restrictions on new offshore production. The result is that what is known as the "West Coast crude surplus" will disappear in 1994 (see Chapter 3), and the region must satisfy a growing oil deficit by importing from abroad. This deficit will increase rapidly into the foreseeable future, thus creating a new oil supply reliability problem for the West Coast as well as expanding the deficit for the nation.

One partial solution to the supply security problem is to resume leasing from potentially large marine resources. These resources are located in federal and state marine lands. This possible solution requires a reexamination of alternative oil resource leasing policies. These important public policy alternatives will be discussed in this chapter. The scope of the analysis will include both federal and State of California policy options.

Even before the Iraqi invasion of Kuwait, there was growing interest in reexamining the system by which natural resources in governmental ownership have been leased to private enterprise for exploration, development, and production. At the federal level, this interest produced the Outer Continental Shelf Lands Act Amendments of 1978. The prior legislation authorized only cash-bonus or royalty bidding, whereas the new legislation specified four alternative bidding systems and then authorized any combination of these bidding arrangements, plus any other system that might occur to the Secretary of the Interior.

The specified bidding alternatives were cash-bonus, royalty, net profit-share, and work-commitment bidding. Each of these bid variables must be paired with one or two required payments which are fixed at the time that bidding takes place. This legislation required experimentation for 5 years with leasing systems other than cash-bonus bidding with a fixed royalty and annual rent payment. After the 5-year experimentation period ended in 1983, Congress allowed the leasing system to revert to its original alternatives. Cash-bonus bidding with a fixed minimum one-sixth royalty and a modest annual rental payment has been and continues to be the primary federal oil and gas leasing system for offshore oil and gas resources. This federal action stimulated the State of Alaska to modify its bidding system and move away from primary reliance on cash-bonus bidding, toward greater use of profit-sharing bidding.

California might follow the federal and Alaskan initiatives by reexamining its leasing system. For this reason, it is appropriate to anticipate the need of the congress and the legislature by providing a review of alternative bid variables as well as fixed-payment requirements that would merit consideration in a revised leasing system.

Following presentation of some background information we will evaluate alternative objectives that might be served by a leasing system. We will then evaluate alternative leasing systems in terms of the major objectives. Finally, alternative leasing policies will be rated in terms of economic efficiency. While it is the responsibility of the U.S. Congress and the California Legislature to set leasing policy, the quality of their work will be improved if voters are well informed about the consequences of alternative policy choices.

BACKGROUND

On the federal Outer Continental Shelf (OCS), 11,513 oil and gas leases covering 59.3 million acres have been issued from 1954 through 1989 (U.S. Department of the Interior, 1990, p. 14). Under a leasing procedure that has become known as the "conventional method," most leases have been issued with cash-bonus bidding and a fixed royalty, usually at 16⅔% of wellhead value. In lease sale # 36 of October 16, 1974, ten tracts were offered under royalty bidding with a fixed bonus of $25 per acre. The royalty bids ranged from 51.8% up to 82.2% of gross wellhead value. Given these very high royalty bids, production should not be expected unless a giant reservoir is found. They should be recognized as low-cost options to produce oil or gas at the option of the lessee.

In addition, under the Outer Continental Shelf Lands Act of 1978, Congress mandated experimental leasing with alternate payment systems for 5 years beginning in 1978. Under this program, 529 tracts were leased, with 314 using a sliding royalty scale, and 215 using a fixed net profit-share payment. The Department of the Interior (DOI) reported that "The Difference in bidding systems relative to their effects on competition, as indicated by the distribution and number of bids per tract by bidding system, is seen to be minimal" (U.S.

Department of the Interior, 1984, p. 34). Further, the DOI found that "Rather than increasing bidding interest by lowering the front-end bonus requirement, as is commonly believed would happen, the higher contingency systems, by making the tracts less valuable tend to result in fewer actual bidders. . . . " (*Ibid.*, p. 39). Based on these findings, the Interior Department has returned to the conventional bonus bid with a fixed royalty. Some use has been made of a one-eighth royalty instead of one-sixth, where expenses are likely to be high due to deep water depth.

From their inception in 1954 through 1989, OCS leases have produced 1.8 billion barrels of oil having a cumulative 36-year market value of $95.8 billion (*Ibid.*, p. 51). In addition, 87 trillion cubic feet of natural gas with a market value of $120 billion have been produced from these leases (*Ibid.*, p. 54). Cumulative revenues received by the federal government from this production is shown in Table 7-1.

California legislation authorizes four alternative bid variables, paired with specified fixed payments at the option of the State Lands Commission as follows. Bidding may be on the basis of (1) a cash-bonus payment in which the Commission must specify a fixed royalty payment; (2) a factor to be multiplied by a specified sliding royalty scale; (3) a flat royalty rate; or (4) a percentage of net profit that will be paid to the State. In all four alternatives, a minimum annual rental payment is required amounting to not less than $1 per acre. Again in all cases, the bid will be awarded to the highest bidder at an oral auction, reserving the right of the Commission to reject any and all bids.

Leasing of California state offshore lands commenced in 1955. The last lease sale was held on August 28, 1968. Planned lease sales originally scheduled for September 1983, were delayed due to jurisdictional disputes between two state agencies, plus political pressure from environmental groups to cease all marine oil development. In the interim between 1955 and 1968, 42 leases have been issued. Production of oil and gas from California state marine lands (within 3 miles of the shoreline) has amounted to 895 million barrels (U.S. Department of the Interior, 1990, p. 82). Revenue from bonus and royalty payments arising out of offshore California state oil and gas leases is shown in Table 7-2.

The record of California leasing has not been subject to a thorough evaluation from an economic perspective. In the absence of such an analysis, we will draw on an extensive review of the federal OCS oil and gas leasing record in

TABLE 7-1   Total Revenue Received by the Federal Government from OCS Oil and Gas Leases, 1954–1989.

Type of Payment	Million Dollars ($ in millions)
Cash Bonus Bids	55,190
Annual Rental Payments	700
Annual Royalty Payments	37,610
Total	93,500

SOURCE: U.S. Department of the Interior (1990), p. 76.

TABLE 7-2   Total Revenue Received by the State of California from State Marine Oil and Gas Leases.

Fiscal Year	Cash Bonus Receipts ($)	Royalty Receipts ($)
1955–56	5,183,643	
1956–57	7,325,774	
1958–59	55,555,774	
1960–61	10,905,111	
1961–62	6,651,635	
1962–63	21,167,654	
1963–64	33,781,572	
1964–65	34,108,012	
1965–66	7,313,193	
1966–67	0	
1967–68	1,635,275	
1968–69	3,101,208	
pre-June 3, 1979		480,554,007
1979–80		41,039,903
1980–81		73,332,309
1981–82		96,769,087
1982–83		83,550,782
1983–84		69,866,000
1984–85		77,600,000
1985–86		58,248,000
1986–87		25,062,000
1987–88		26,438,000
1988–89		20,375,000
1989–90		16,813,000
Total	186,728,851	1,069,648,088

SOURCE: State of California, State Lands Commission, *Annual Report of the State Oil and Gas Supervisor*, varous issues.

the Gulf of Mexico. This examination and analysis included 1,233 leases issued over the years from 1954 through 1969, together with the record of production from leases through the year 1979. Beyond 1979, projections were made of lease production and revenue through lease close-down, not later than the year 2010.[1]

It should be made clear that when a firm purchases a prospective oil or gas lease it buys not one but two assets: It buys the obvious right to explore for and produce any hydrocarbons found on the lease; and, less obviously, it buys information not only for the tract acquired but for adjacent tracts as well. Prior to bidding, firms will normally conduct or otherwise obtain geological and geophysical data which might identify prospective hydrocarbon-bearing structures covering specific leases.

Figure 7-1 shows two such potential structures centered on Tract E. The firm that obtained Tract E in a lease sale may have identified two potential structures either before or after winning the tract. Drilling at the site might have established that the smaller structure was dry and as a consequence the lessee has obtained valuable information indicating that unleased tracts B and C are of little or no value. Hence, in any subsequent sale, the lessee of Tract E will use his information to avoid overbidding on tracts B and C. The larger of the two potential structures may, on the other hand, have shown promising prospects on the basis of post-sale drilling. In this event, the lessee of Tract E could use his superior information to bid more wisely (more aggressively) on any subsequent lease-sale offering of tracts D and F.

Evidence in support of the above points is available from rate-of-return analyses of Gulf of Mexico OCS lease-sale returns. Table 7-3 shows that for all 1,223 leases issued in the Gulf of Mexico OCS lease sales occurring from 1954 through 1969, lessees in the aggregate earned a 10.74% average after-tax return on their total lease investments.[2] On the 1,109 *wildcat* leases issued in this group, lessees earned a relatively low 10.04% return. However, on 63 *drainage* leases won by firms that owned all or part of an adjacent tract sharing the same hydrocarbon-bearing structure, lessees earned a 16.20% rate of return.

---

[1] This paper will draw heavily from reports by Mead, Moseidjord, and Sorensen (1983); Mead and Pickett (1984); and Mead and Moseidjord (1984).

[2] This rate of return includes actual oil and gas production through 1979, plus forecast production and value through well abandonment not later than the year 2010.

These results indicate that for wildcat leases, the observed low rate of return is consistent with the hypothesis that high bids were too high relative to the hydrocarbons produced from such leases and that, if the value of information obtained from such wildcat leases was added to the observed rate of return, the results would more closely correspond with a competitive norm. At the same time, "neighbor firms" (firms that owned part or all of the productive wildcat tracts adjacent to the drainage tract and sharing the same structure) earned a super-normal rate of return. These results are consistent with the hypothesis that information acquired and paid for as part of adjacent wildcat leases yielded compensating gains when neighbor firms acquired adjacent drainage tracts.

TABLE 7-3   After-tax Rate of Return by Lease Class.

Lease Class	After-tax Rate of Return[a] (%)
All 1,223 leases	10.74
1,109 wildcat leases	10.04
63 drainage leases won by neighbor lessees	16.20

SOURCE: Mead, Moseidjord, and Sorensen (1983), p. 41.
[a] The first year of OCS lease sales, 1954, is treated as year zero. Cash flows are aggregated by calendar year and discounted to 1954 regardless of the year of the lease sale.

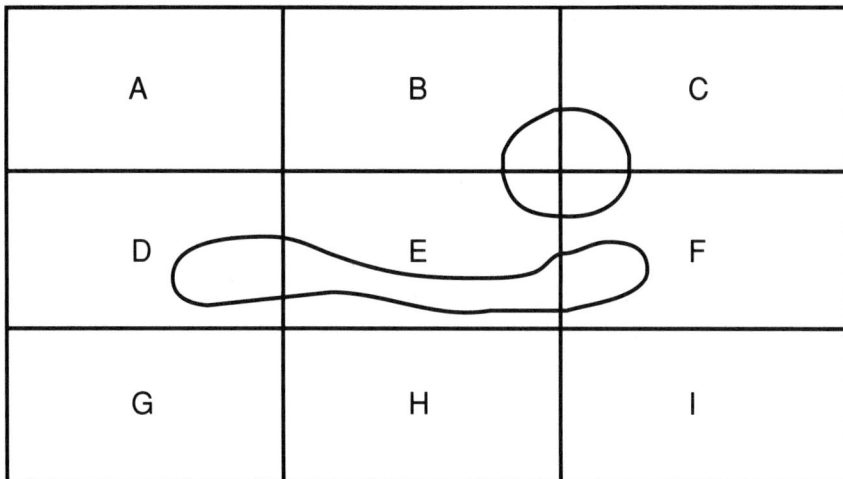

FIGURE 7-1   Diagrammatic Illustration of Potential Hydrocarbon-bearing Stuctures Crossing Several Lease Tracts.

ALTERNATIVE OBJECTIVES OF A LEASING SYSTEM

The primary purpose of the remainder of this chapter is to analyze the four major alternative bid variables: (1) cash bonus; (2) royalty; (3) net profit-share; and (4) a work-program to be performed by the lessee, and a "threshold" rate used in conjunction with a Resource Rent Tax (RRT) payment system. Subsequently, we will consider fixed factors, including all of the above, plus a rental requirement and a resource rent tax. This analysis requires a prior statement of objectives that a leasing policy should achieve.

Maximize and collect all of the economic rent

As a prerequisite to this evaluation, policy goals must be identified as clearly as possible. There is wide agreement among natural resource economists that a single objective of "optimum resource allocation" should guide resource management policy. A statement of that objective has been carefully and concisely stated by McDonald as follows:

> The Department of the Interior, as custodian of the affected federal lands, should lease lands for minerals production under such terms and conditions, and at such a rate, as will tend to maximize the present value of the pure economic rent derivable from them.

McDonald then defines economic rent as:

> any surplus in the income of a factor of production—land, labor, or capital— where *surplus* is the excess of the factor's income over the minimum amount necessary to call forth its productive services. . . . Clearly, such a surplus could arise only where the supply of the factor of production is less than perfectly elastic; for if the supply were perfectly elastic, the quantity supplied would expand in response to the least surplus and immediately wipe it out (McDonald, 1979).

While McDonald presents his statement of objective in terms of federal government leasing policy, the statement applies equally well to oil and gas resources owned by the State of California. Firms that maximize profit over time also maximize the present value of their resources, and society's interest will be served subject to the conditions that: one, no monopoly power is exercised; two, either no externalities exist or they have been effectively internalized; and three, other government interference does not distort prices or resource allocation. These same conditions also apply to government management of resources.

The concept of economic rent is easily illustrated in Figure 7-2. The height of the bar represents the total revenue obtainable, subject to a cost constraint, from any given oil or gas lease. This gross revenue is reduced by the necessary costs of exploration and production shown in the lower segment of the bar. Under conditions of effective competition, the operator is forced by the market to avoid unnecessary costs over which he exercises control. Government regulations that impose social costs in excess of social benefits on lease operators should be avoided. Such constraints either reduce production and revenue (lower the height of the bar), or impose unnecessary costs on the operator. In either case, the residual economic rent is reduced and society suffers a loss. Public policies that result in social costs greater than social benefit are not consistent with resource conservation goals.

The future revenue stream, economic rent, and all necessary costs are represented in Figure 7-2 to be in present value terms. Given the assumptions above regarding competition and externalities, this rate corresponds with the discount rate that a firm would use in computing its optimal bid value. The

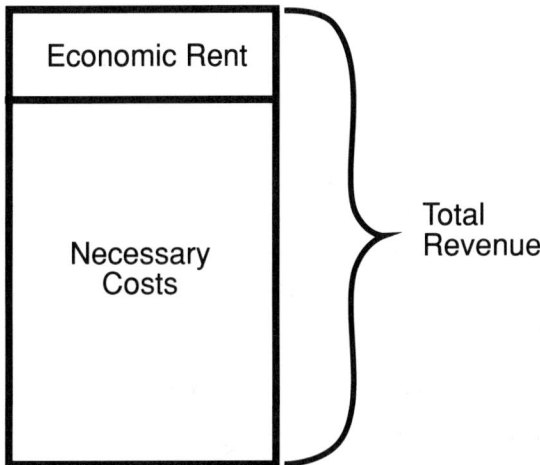

FIGURE 7-2   Model of Economic Rent Estimation. All values are discounted to obtain net present values. Economic rent is not a return on invested capital; the latter is part of the "necessary costs" of production. Necessary costs exclude payments to government. (The normal return on the lessee investment is provided in the discount rate.) Furthermore, this model applies to economic rent estimation in the aggregate, where productive leases generate sufficient revenue to cover dry holes and tracts that are productive but fail to generate revenue equal to their necessary costs.

discount rate includes a normal competitive risk-adjusted rate of return on the capital employed.

Economic rent is paid in several forms. (1) Specified annual rent payments are commonly required. These payments cease when production begins. (2) Royalty payments normally amounting to one-sixth (16⅔%) of the wellhead value of any oil and gas produced are required. (3) The discounted present value of the remaining economic rent is theoretically collected in the form of the cash bonus or other bid variable. The leasing method used affects the amount of economic rent available for collection because the leasing method may affect the future revenue stream and the amount of the exploration, development and administrative costs imposed on the lessee.

Unambiguously identify the high-bidder

One purpose of auction bidding for leases is to determine in an "arm's-length" manner what firm is to be awarded the lease. Resource conservation goals are served only if economic forces are present which tend to allocate leases to the most efficient operators. A bidding system that allows the bidder to bid simultaneously on two variables creates the "apples and oranges" problem—there is no objective means of determining which firm is the high bidder. For example, under the State of Louisiana bidding rules a bidder may make a high bid on the bonus payment, paired with a low bid on the royalty. Another bidder may do the opposite. Without knowing how much oil and gas will be produced and when, there is no way to identify the high bidder. Even when a single bid variable is mandated, if the rules allow renegotiating after the award has been made, the apparent initial high bid may turn out to be an illusion. This problem occurs in both the Australian work-program system and the U.S. government royalty bidding system.

Minimize administrative costs (relative to benefits)

The bidding and payment system in combination should minimize the costs of administering the system by which the government collects its economic rents. A system that imposes unnecessary accounting, supervising, negotiating and litigating costs on the lessee reduces the available economic rents. This can be easily seen in Figure 7-2. As the expected costs of lease administration paid by the lessee increase, such increases will be deducted from expected revenues to obtain the economic rent estimate. Since the present value of the expected economic rent corresponds to the bid amount in a pure bonus-bidding system, the government ultimately bears the expected lessee cost of lease administration.

This problem would be included in the section above (dealing with maximizing economic rent) except for the fact that administration costs borne by the government (the lessor) have no direct effect on economic rents. Rather, unnecessary costs of government lease administration may be thought of as an unnecessary expenditure of the economic rent that the government has collected. Thus, unnecessary administration costs constitute either a reduction in economic rent, or an unnecessary use of the collected rents.

Equate the present value of lessee payment to the government to the present value of the actual economic rent

Drilling into a potential oil-bearing structure is not allowed in either federal or State of California leasing procedures. Interested potential bidders are limited to information gained from geophysical and geological research, plus any information from operations on adjacent tracts. Consequently, it is impossible for any bidder to know with precision whether a promising structure contains any oil or gas, or if it does, how much, and what its quality might be. Some critics have charged that bidding for leases "is a crap game" where only the laws of chance govern. But as will be shown later, in the aggregate there is a correlation between bonus bids for leases and the value of oil and gas produced from leases.

Because of this uncertainty, risk of great gains or losses on individual leases must be assumed by either or both the lessor and lessee. If lump-sum payments are made in advance of drilling and production, risk is borne heavily by the lessee. This risk may be shifted to the lessor by means of a payment system that is a function of the volume or value of the oil and gas produced, or even of net profit (before any lease payments). These systems have been called "contingency payment systems."

Ideally, payments for a lease should correspond closely with the value of the economic rent. However, the systems that have been developed to attain this correspondence have involved social costs due to inherent inefficiencies of the system. We will evaluate alternative bidding and payment systems not only in terms of their harmony between "what you pay is what you get," but also in terms of the economic efficiencies that are part of the trade-off between lump-sum and contingency payment systems.

Aid small business

Similarly, one finds frequent reference to aiding small business as an objective of leasing policy. From a resource conservation perspective, leases should be awarded to the most efficient firms. The high-bid firms tend in the long run to be the most efficient firms. If legislation biases lease allocation in favor of small firms that are less efficient (indicated by an inability to compete), then scarce oil and gas resources are not produced effectively and the result will be a slower economic growth rate, an inefficient allocation of scarce resources (the opposite of resource conservation), and a lower lease income level than would otherwise exist. It should be noted that there is no evidence indicating that the large multinational oil companies are more efficient operators in the OCS environment than are smaller firms. To the contrary, the record of federal leasing shows for that for the big-eight international oil companies more of their acquired leases were dry, and they earned a lower rate of return on their lease investments compared to the smallest firms. This record is shown in Table 7-4.

The issue of technical competence must be considered prior to introducing any small-firm inducements to enter the marine oil and gas lease sale market. Exploration for and production of oil or gas from the marine environment requires a high degree of technical expertise, because a mistake can impose significant costs on society. If technical competence is positively correlated with firm size, then the argument for selecting a leasing system which facilitates entry by small firms is flawed. While one may offer a hypothesis reflecting the foregoing position, such a hypothesis has never been tested.

Further, a rarely discussed issue relative to small firms concerns reliability or responsibility. Would society's interest in environmental protection be

TABLE 7-4   Results of Lease Operation for 1,223 OCS Oil and Gas Leases, Gulf of Mexico, Issued from 1954 through 1969.

	Average Bonus per Lease ($)	Aggregate Internal Rate of Return After Taxes (%)	Dry Leases as Percent of Total Leases
Big-8 Firms	2,310,499	10.37	61.70
Big 9–20 Firms	2,354,070	11.26	63.71
All other lessees	1,740,377	11.15	57.90
Average, all firms	2,228,332	10.74	61.57

SOURCE: Mead, Moseidjord, and Sorensen (1983), p. 41.

equally served by small firms as lease owners and operators? Large and small firms differ in many respects. One difference is based on the lifetime concept of the firm ownership and management. Large firms act as if they have an infinite life, whereas some small firms have a more limited perspective. This leads to the possibility that small firms, when confronted with a potentially expensive oil well blowout, might elect to "walk away" from a lease. An expensive blowout in the marine environment could easily involve an uninsured cost of suppression and cleanup, plus legal liability for damages, that would exceed the net worth of the small firm, a situation very unlikely for a large oil company. This point has been demonstrated in the case of the *Exxon Valdez* tanker oil spill. The most spectacular oil well blowout and consequent oil spill in U.S. history of production from the marine environment occurred in the Santa Barbara Channel on January 24, 1969. Would society have been better off relative to environmental protection if that lease had been owned by a small firm rather than by Union Oil Company (as the operator), Texaco, Gulf and Mobil? If the answer to this difficult question is "worse," then a policy specifically favoring entry of small firms in the marine lease market is of questionable social value.

Finally, small firms may suffer scale-economy disadvantages that do not allow them to successfully compete in the high-cost marine lease market. Wilson has argued that the smaller firm's inability to bear the risk as easily as a larger firm does is due to its limited capital and to its poorly diversified portfolio of leases. In fact, failure to find sufficient reserves on just a few leased tracts can jeopardize a small firm's entire existence. For this same reason a small firm's access to capital markets may be severely limited or exorbitantly expensive. Potential lenders or investors in a small firm recognize that the use of their funds for a cash bonus provides no security that the funds can be repaid or that a normal rate of return will be earned (Wilson, 1978, p. 637). This is an economy-of-scale problem. It suggests that large firms are more efficient in this high-risk, high-capital requirement situation. This economic fact of life may be accepted for its public policy implications, or resisted with policies that subsidize small-firm entry. Wilson argues for a two-phase leasing system in which exploration leases would be marketed separately from production leases. He notes, however, that a two-stage system creates new difficulties and he is unable to recommend an economically efficient solution.

Stimulate competition

Another objective frequently voiced is that leasing policy should stimulate competition. Competition is essential if lease-sale markets are to perform effectively. Steps might be taken to increase the number of competitors able to participate in the lease auction market and thereby increase competition for leases. These steps involve primarily reducing barriers to entry in the form of front-end payments to the government. Whether these steps are appropriate depends on whether competition is effective under the present leasing system.

From 1954 through 1981, when 19,074 tracts were offered for lease by the federal government in the Gulf of Mexico, the average number of bids received per tract was 3.2. Of the tracts offered, 47% received no bids. The government reserves the right to reject any and all bids and exercises this right when it believes that the high bid received is inadequate, or for any other reason. The record shows that from 1954 through 1981, high bids were rejected by the government on 12% of the tracts receiving bids.

From the above information, one might erroneously conclude that competition is inadequate inasmuch as about half of the tracts offered received no bids and 12% of the bids offered were rejected. However, the record of 1,223 tracts actually leased through 1969 shows that 61.9% of those tracts were dry. Another 16.3% did not produce enough revenue to cover their costs. Thus the tracts receiving no bids were probably of no net value in the aggregate. It is also likely that at least 80% of the tracts where bids were rejected by the government were of negative value (the *in situ* value of any reserves was negative). Supporting these generalizations that bidders in the aggregate are able to distinguish productive from nonproductive tracts is the finding from a multiple regression analysis that there is a positive and significant correlation between high bids, as a dependent variable, and the value of oil and gas actually produced from tracts leased (Mead, Moseidjord, Muraoka, and Sorensen, 1985, pp. 65–77).

Additional evidence on the basic rationality of the bidding process is shown in Table 7-5. Tracts receiving very low bonus bids have a very high percentage of dry leases and a very low value of production. Correspondingly, tracts receiving very high bonus bids have a low percentage of dry leases and a high value of production.

Evidence of effective competition for leases actually issued under cash-bonus bidding was given earlier in Table 7-4, showing that the nominal after-tax rate of return on such leases was 10.74%. This is clearly not in excess of the average nominal 11.8% earnings after taxes for all U.S. manufacturing companies over the same time period. If one uses a 12.5% nominal discount rate, the present value per tract leased after taxes was negative $192,128 (Mead, Moseidjord, and Sorensen, 1983, p. 41). Thus, even under a bonus-bidding system which requires a "front end" payment of economic rent not collected as annual rent or royalty payments, the evidence shows that competition for leases is effective. This result casts doubt on the argument that additional measures are needed in order to increase competition for leases.

In sum, the evidence indicates that lease-sale markets are effectively competitive. Additional steps that might be taken to increase competition for leases need to be evaluated relative to the costs of such policies. Political discussion concerning leasing policies that tend to lower barriers to entry and thereby increase competition is often in terms which indicate a second interest—aiding small business. Given effective competition, the issue of subsidizing small business should be debated separately. The latter issue raises a question of whether the public interest would be served by the entry of smaller firms into OCS oil and gas exploration, drilling, and production. The record of bidding for OCS leases shows that in the 20-year period from 1954 through 1973, 128 separate firms have won OCS leases either as solo bidders or as part of a bidding combine (Wilcox, 1975, p. 92). Through the year 1969, the big-20 firms won 84% of the leases issued. The remaining 16% of the leases were won by firms that range widely in size below the big-20 group.

TABLE 7-5   Relationship of Bid Prices to Production Results.

Bonus Bid Class	Number of Leases Issued	Percent Dry Leases	Average Bonus per Lease ($)	Undiscounted Average Gross Value of Production per Lease, Actual through 1979 ($)
$250,000 or less	354	81.4	126,450	4,996,521
$250,000–1,000,000	367	64.6	524,998	9,660,978
$1,000,000–3,250,000	285	48.8	1,874,621	20,807,215
More than $3,250,000	217	41.0	9,002,511	42,295,839

SOURCE: Mead, Moseidjord, and Sorensen (1983), p. 41.

Miscellaneous economic issues in leasing policy

The nonrenewable nature of oil and gas. Oil and gas resources are non-renewable natural resources. Consequently they must be allocated over time in a manner that allows their price to perform a conservation function so that, as economic exhaustion is approached, substitute and higher-cost energy sources such as solar, oil shale, and possibly nuclear fusion become economically producible. The operating rule is that optimum resource allocation requires that the *in situ* net value of oil and gas increase at a compound interest rate corresponding with the opportunity cost of capital. An equivalent and alternative rule is that the present value of resources should be maximized.

The record of federal and state stewardship with respect to optimum resource allocation over time is not satisfactory when judged by a resource conservation standard.[3] There is a governmental bias favoring "rapid development" ("due diligence") of oil and gas reserves in public stewardship. But there is no study within the California or federal governments addressed to the question of the optimum allocation of resources between present and future uses. The long-standing "5-year rule" at the federal level requires that lessees begin producing oil or gas from federal leases within a 5-year term or forfeit their lease to the government. California statute allows only 3 years to begin drilling. Failure to meet this requirement causes the lease to terminate. Subsequent failure to commence operations and "prosecute them diligently" again leads to lease termination. Thus any lessee who is concerned that oil and gas resources are being depleted too rapidly with the result that the next generation will suffer a severe decline in supplies prior to the availability of substitute resources is not permitted to delay production. Such diligence rules force current production independent of future supply-and-demand prospects and probable price behavior. Thus, "speedy development" should be rejected as an operating objective for federal lease management (Mead and Muraoka, 1987).

From the point of view of maximizing the present value of revenue flowing to the government, short lease terms should be avoided. If bidders for leases believe that prescribing early production reduces the present value of the potential net income flow from a lease, perhaps because they believe that oil and/or gas prices will be substantially higher in the future or costs will be sub-

---

[3] The resource conservation standard generally accepted in public discussion is equivalent to the optimum resource allocation (Pareto optimality) standard in the professional economics literature.

stantially lower, bids that are offered will be correspondingly reduced. Thus, lease income to the government will be reduced without any obvious offsetting gains.

Confusing leasing objectives. The public is occasionally given political statements of leasing objectives which are virtually meaningless. For example, a recent U.S. House of Representatives' report specified that the bidding regime should "strike a proper balance between securing a fair market return to the Federal Government for the leasing of its lands, increasing competition in the exploitation of resources, and providing the incentive of a fair profit to the oil companies, which must risk their investment capital" (U.S. House of Representatives, 1977, p. 64). This statement of objectives will generate no disagreement, but it is meaningless insomuch as "proper balance," "fair market return," and "fair profit" are undefined.

Tradeoffs in bidding. It is important to understand that tradeoffs exist between amounts bid for the bid variable and the fixed required payments to the government. For example, California lease-sale contracts require payment of at least $1 annual rent per acre. The actual payment is fixed by the State Lands Commission in advance of bidding. Where the bid variable is a cash bonus, there will be approximately a dollar-for-dollar tradeoff in terms of present values such that the higher the required rent payment, the lower will be the bonus bid.[4]

However, in another tradeoff setting, one should not expect a dollar-for-dollar relationship. For example, as we will see later, royalty payments lead to reduced production of oil and gas from leases. Consequently, where the government doubles its ordinary one-sixth royalty to a fixed one-third royalty, total lease revenue will decline as shown in figures 7-2, 7-3a and 7-3b. High royalty rates (a tax on *gross* revenue) leads to premature abandonment of leases

---

[4] The only allocative efficiency consequence of annual lease rental payments is that it leads to earlier production than is optimal. The rental payment can be avoided or minimized by prompt exploration and immediate production. Current federal and California regulations cause any rental payments to cease as soon as royalty payments begin. The optimum time profile of production would occur without rental and royalty payment requirements and with repeal of the 5-year rule. Lessees would the be free to maximize the present value of their leasehold. Thus, they would choose between present and future production such that the value of their leasehold would increase with the opportunity cost of capital. Well-established economic theory indicates that choices made under this rule will also increase the general welfare if there are no significant externalities and the opportunity cost of capital corresponds with the social time preference rate.

and therefore reduced oil and gas production. On a present-value basis, bidders will reduce their bonus payments by more than a dollar-for-dollar increase in estimated royalty payments. Consequently, high fixed or sliding-scale royalty rates should be avoided.

It should also be reaffirmed that any additional costs that the government imposes on lessees are paid dollar-for-dollar in terms of present value foregone by the government. That is, in calculating the net present value of tracts in order to submit bids, bidders subtract from expected revenues all expected costs of exploration, development, and production. For example, if lessees are required to spend $1 million for an environmental impact statement and are aware of this requirement prior to bidding, they will deduct the present value of this payment from their computation of the present value of the lease. Thus, the citizens of the federal or state governments pay for such requirements. If inefficient methods of exploring for, producing, storing or transporting oil or gas are required by federal, state or county regulations, the present value of the costs (or probable costs) will be deducted from the present value of the bid variable.

When such requirements are imposed on lessees after bidding has been completed, the *money costs* are borne by lessees, except for the consequent reduction in profit and therefore in income tax payments. However, lessees learn by experience; in subsequent bidding for leases, they will estimate the probability of repeat regulation and subtract the present value of such probable costs from their bids.

Apart from the *money costs*, use of resources (labor, steel, etc.) regulations is always a *social cost* and is borne by society at large, since these resources cannot be used on other worthwhile projects. This is an important point for government officials and the public to understand because the common incorrect understanding is that costs imposed on oil company lessees rest with the lessee.

Finally, any additional uncertainty concerning the terms of a lease sale and rules and regulations that may be imposed in the future by the government, leads bidders to lower the expected present value of the lease and reduce their bids. Uncertainty will be reflected either in higher discount rates or in assumed additional costs. Either avenue leads to lower bids. The terms of the lease should be clearly specified in the lease contract prior to bidding in order to reduce uncertainty and the consequent loss in present value when bids are calculated and submitted. Changing the rules of the game after the contract has been concluded should be avoided because it leads to increased uncertainty, higher discount rates, and lower present values in future lease auctions.

Summarizing this section, we find that the first three objectives of a leasing system stand out as of great importance relative to choice of a bid variable: (1) maximizing and collecting all of the economic rent; (2) unambiguously identifying the high-bidder; and (3) minimizing administrative costs.

The fourth part of this section, equating the present value of payments to the government to the actual economic rent due to the government, is of great importance in the aggregate. However, for firms that have a diversified portfolio of leases, equating these values on a lease-by-lease basis is of questionable value, and is probably not obtainable without accepting offsetting social costs.

The next two parts, concerned with aiding small business and stimulating competition, appear to be minor considerations with strong political overtones and relatively little economic content.

ALTERNATIVE BID VARIABLES

We will now turn to a systematic evaluation of alternative bid variables, followed by a discussion of the fixed-payment alternative in the next section. The analysis will be in terms of the principal performance criteria as discussed in the preceding section.

In the bid calculation process, both revenue and costs are estimated in advance of bidding by potential bidders. Expected dry tracts, or tracts expected to have costs in excess of revenue (including option value and the value of information about adjacent tracts as implicit revenue), receive no bids. If competition is effective, bidders in the long run will be forced to bid away all of the economic rent. However, the bidding system, together with any fixed payment and regulations, has an important effect on both production (and therefore revenue) and costs, hence on the amount of economic rent available and collected by the government.

Pure cash-bonus bidding

A major advantage of a pure cash-bonus bidding system is that it transfers all economic rent to the government at the time of the auction. Upon payment of the bonus, the economic rent becomes a "sunk cost" and thereafter has no effect on investment or output decisions. These decisions will then be made on the basis of incremental private costs and benefits. This result is optimal because economic rent is not a social cost. Rather it is a transfer payment to the government and as such it should not be allowed to affect investment or output decisions.

Several disadvantages have been charged against bonus bidding. For one, this system requires a front-end payment, thus discouraging small-firm (low capitalization) entry. Whether this consequence is serious depends on whether additional competition is needed from small firms that suffer diseconomies by reason of inadequate capitalization. Judgment on this issue rests partly on the environmental implications of small-firm operation in an environment where operator technical competence and responsibility are desirable attributes. This issue was discussed above. It also rests partly on the competition issue. Reducing or eliminating front-end payments would tend to increase the number of firms competing for offshore oil and gas leases. This would be an important item if competition for such leases is inadequate. Data on the aggregate return earned on federal oil and gas leases in the Gulf of Mexico indicate competitive results, as shown in Table 7-3. For the 1,233 leases studied, the average number of bidders was 3.33 per tract leased.

In addition, the front-end payment characteristic also has been criticized because it requires that the net value of the future economic rents (not collected as royalties, annual rents, and taxes) be discounted to be paid as a cash bonus. If the lessee is risk-averse relative to the lessor, then the front-end payment is less than the present value of the lease to the government. Leland wrote that

> Excessive risk aversion by firms incurs social costs . . . [leading] to over-discounting for risk and, therefore, to lower bids on tracts. The government will get less economic rent than it would if socially efficient production and bidding decisions were made" (Leland, 1978).

This viewpoint has been advanced by economists evaluating alternative leasing systems for the U.S. government, the State of Alaska, and Australia (Leland, Norgaard, and Pearson, 1974; Emerson and Garnaut, 1984; Gaffney, 1977).

The Leland conclusion rests on an assumption that bidders are risk-averse and therefore require a risk premium. Two objections should be made to the Leland point. First, while it is true that lease acquisition (including both the exploration and development—production phases) involves extraordinary risk, portfolio diversification can convert such investments into a normal rate-of-return situation. In the Gulf of Mexico, the 20 largest lessees (rated according to number of leases held) acquired 93% of the leases issued over the years 1954 through 1969. The firm having the smallest number of leases among the big-20 had 12.5 leases (including joint venture shares). Because smaller firms commonly bid through joint bidding ventures, diversification is broader than these numbers suggest.

Second, the assumption of high risk aversion was not empirically tested by any of the authors noted above. Empirical research has indicated that rates of return for federal OCS oil and gas leases do not reflect returns above the normal for all U.S. manufacturing firms. If anything, oil and gas rates of return are subnormal. The average after-tax rate of return on equity capital earned by all manufacturing firms in 1954–1980 was 11.8%. The comparable rate of return on 1,233 OCS oil and gas leases was 10.74%. This record shows no evidence of a market risk premium in oil and gas leasing. Nearly all firms participating in the lease market hold a diversified portfolio of investments in many oil and gas leases and in a variety of industries.

A third disadvantage of the bonus-bidding system appears to have some potential merit. The Australian mining industry has pointed out that what is called "sovereign risk" is a hazard of bonus bidding. This term is defined as "the risk that the fiscal arrangements will be changed by governments after the allocation of mining rights" (Emerson and Garnaut, 1984, p. 134). When potential bidders calculate their bids, they forecast future revenue and cost streams on the basis of existing rules, perhaps adjusted for the probability of unfavorable changes in those rules.

One of the rights of a sovereign government is to change the rules (increase taxes, royalties, environmental regulations, etc.) after bids have been submitted. The U.S. government exercised this right when it added costly environmental regulations on existing oil exploration and development leases following the Santa Barbara oil spill in 1969, increased tax payments under the Windfall Profits Tax legislation in 1980 and again imposed new shipping and other oil-spill environmental regulations after the 1989 *Exxon Valdez* oil spill. Of a more serious nature, the State of Alaska increased severance taxes on production from the Prudhoe Bay field following discovery of that super-giant reservoir.

When unanticipated new costs are levied against lease production after bids have been submitted, the bonus payment will be too high; but it has already been submitted and accepted. When governments legally violate agreements in this manner, firms learn by experience and in subsequent bids introduce a probability estimate of future cost increases due to the sovereign risk. If the estimated cost does not materialize, the bids will fall short of reflecting the true value of the property. If a cost is imposed and is fully anticipated and discounted, then the government bears the full cost of changing the rules. The only disadvantage attributable to bonus bidding, therefore, is the possibility that bidders will systematically overestimate the probability that the government will impose post-bid costs. This does not appear to be a significant disadvantage, and may be more than offset by advantages of the bonus-bidding system. We have no empirical data to support or deny this point.

From the viewpoint of keeping administrative costs at a minimum, cash-bonus bidding is preferred by a wide margin. With this system, the winning bidder pays for his lease with a lump-sum payment and no further payments are required. There are no lease terms for the government to monitor. The lessee is entitled to produce any oil or gas found on the lease. An optimizing lessee will produce from the lease, provided that the present value of production exceeds the present value of the additional costs. He will cease production when incremental recovery costs reach the incremental value of the product. Administrative costs for both lessee and lessor are at a minimum. In the absence of any externalities, this private judgment corresponds with the social optimum.

While the high bidder is unambiguously identified in pure bonus, royalty, and profit-share bidding, only in the pure bonus bidding case is there a theoretical basis for asserting that, in the long run, high bidders are the most efficient operators. In this instance, the high bidder indicates his expectation that costs are low relative to the expected income flow. This expectation in the long run must be correct or the firm will face losses in its lease development operations. The bonus is paid in advance, and the government knows exactly what its revenues will be when it accepts the high bid.

Pure royalty bidding

In the case of royalty bidding, the loss of output and economic rent can be extremely large, relative to bonus or profit-share bidding. The loss follows from the facts that a royalty payment is a percentage of *gross* income, not *net* income, and decision-making is distorted relative to bonus payments. Three negative consequences follow for the lessor's collection of the economic rent via royalty payments.

To begin, royalty payments levied on each unit of production are added to other incremental costs of production. Consequently, they lead to early abandonment of production. Because royalty payments are transfer payments and not real social costs, *premature* abandonment occurs. This point was illustrated in figures 7-3a and 7-3b. The production history of a typical oil or gas well is shown in Figure 7-3a. Production begins at a high level and proceeds along some exponential decline rate as reservoir pressure diminishes. As production falls, incremental costs of production per barrel increase. Figure 7-3b shows that, without a royalty payment, production will cease at point T4 when incremental costs rise to equality with the price of oil. This point is a societal and a private optimum—with no obvious externalities and a competitive domestic oil market.

However, with the conventional State of California and federal OCS minimum 16⅔% fixed royalty requirement, production would be abandoned at point T3 when incremental costs, including the royalty payments, rise to equality with the price of oil. Even with this relatively low royalty, a well will be prematurely abandoned, leaving socially valuable oil or gas in the ground equal to the output from T3 to T4.

In federal OCS lease sale # 36 on October 16, 1974, ten tracts were offered under royalty bidding with a fixed $25 per acre bonus. Bids were received on eight of these tracts. The royalty bids ranged from 51.8% up to 82.2%. We have plotted royalty payment of 52% and 82% in Figure 7-3b to illustrate the approximate magnitude of the premature abandonment problem under high royalty requirements. We find that when incremental costs amounting to 52% of gross wellhead value are added to the socially necessary incremental lifting costs, abandonment occurs at T2. With an 82% royalty requirement, abandonment would occur at T1 leaving a large amount of valuable oil in the ground. It is true that the oil will be there for subsequent development and production, but again, once a field is abandoned, reopening it at a later date requires new incremental costs in the form of drilling and development that would have been unnecessary if the initial production had continued.

Production (bbl. of oil equivalent)

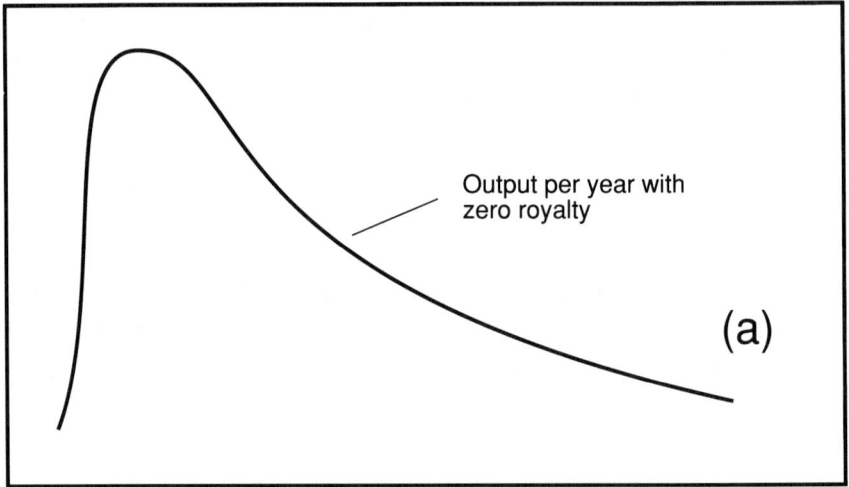

Output per year with
zero royalty

(a)

T4

Time

Incremental Cost ($/bbl. of oil equivalent)

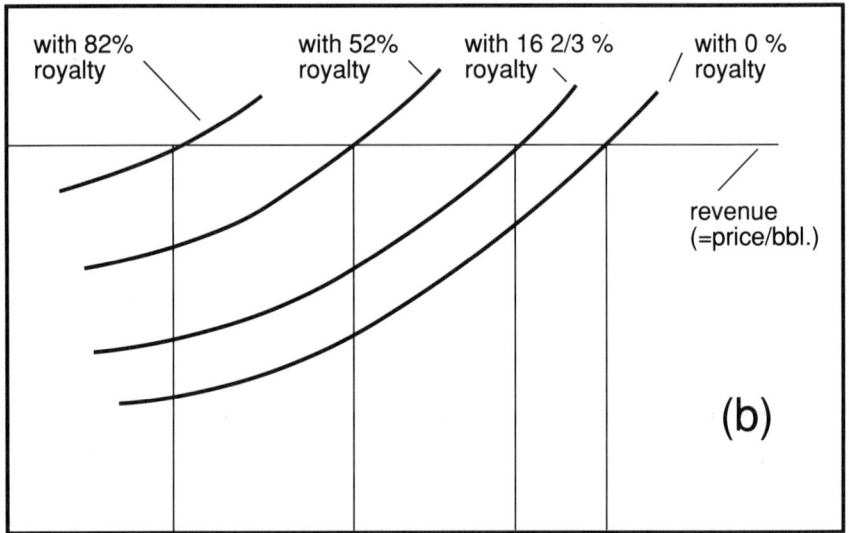

with 82%
royalty

with 52%
royalty

with 16 2/3 %
royalty

with 0 %
royalty

revenue
(=price/bbl.)

(b)

T1   T2   T3   T4

Time

FIGURE 7-3   (a) Diagrammatic Illustration of Oil Well Productivity over Time, (b)
and Incremental Cost and Revenue Flows over Time.

Congress gave the Secretary of the Interior authority to reduce the fixed royalty rate down to zero in order to avoid the premature abandonment problem. However, no Interior Secretary has ever exercised this authority. The primary consideration here appears to be a political hazard, that is, the public might not understand the economic reasoning indicating that society would benefit by such action and might instead see it as a "give away" to oil companies. Further, current legislation allows the Interior Secretary to set the fixed royalty rate as low as $12\frac{1}{2}\%$, thereby reducing the premature abandonment loss. However, this right has been exercised only experimentally in the years 1978–1983. Similarly, California Statute (PRC 6827.2) authorizes the State Lands Commission to reduce the minimum royalty in order to prevent premature abandonment of a lease. Also, similarly, this provision has never been used by the Commission.

A second negative consequence follows from the fact that such incremental outlays as enhanced oil recovery investments (EOR) add to both operating costs and revenue, but royalty payments are a levy on gross, not net revenue. Many such investments will become uneconomic as a consequence. Thus, royalty payments lead to suboptimal levels of investment in reservoir development.

Finally, the higher the royalty requirement, the greater will be the discoveries that will be plugged and abandoned rather than developed. One can see from Figure 7-3b that incremental profits per barrel are very small for high royalty levels. The incremental costs of production do not include the cost of field development. Hence only giant fields will be produced under the type of royalty bidding that produces relatively high royalty rates.[5]

In all three instances, the collection of economic rents will be reduced under royalty bidding. This is a major negative factor relative to cash-bonus bidding. Reference back to Figure 7-2 shows that as total production and total revenue is reduced by the three factors discussed above, the amount of economic rent from oil and gas resources also declines.

---

[5] While Leland ignored the third misallocation listed above, he reached the same conclusion based on a mathematical proof. He wrote as follows: "Royalties will *always* induce inefficient production decisions by firms. The 'early shut-down' problem is but one aspect of this result. In general, investment in exploration and development will be less as well" (Leland 1978, p. 432).

As a variation on the royalty bidding system outlined above, both the State of California and the federal government have authorized and experimented with sliding-scale royalty payments where bidding is on some factor equal to or greater than 1.0 times a specified sliding scale. A sliding scale would be an improvement over an inflexible royalty rate if the low-end rates are below the conventional 16⅔% fixed rate, but this system creates a new set of problems. Lower rates must be "triggered into effect" by some objective measure. This measure is commonly the output per well, over some time period.

Two problems appear in sliding-scale systems. First, the operator is in control of output and can reduce this payment to the government by avoiding high output levels, thereby stretching output over a longer period of time. But a social optimum requires that the output profile be such as to maximize the present value of the resource independent of economic rent transfer considerations. This problem is a minor one because some other producers who are free to set their annual output levels so that the value of oil in the ground increases at the opportunity cost of capital (the condition which maximizes the present value of oil reserves) may set their annual output levels in an offsetting manner.

Second, production-enhancing investments are discouraged by the fact that increased production may escalate royalty rates into higher brackets on the sliding scale and some (perhaps all) of the incremental benefits will be taken by government.

In sum, royalty bidding, even if modified and improved by a sliding scale, reduces the recoverable oil and gas from productive fields. In terms of Figure 7-2, total revenue is reduced, without corresponding reductions in costs, with the result that the nation or state loses economic rent. Royalty bidding should be rejected on the basis of its premature abandonment problem and consequent resource misallocation effect. A fixed royalty rate should be avoided for the same reasons.

From the viewpoint of minimizing administration costs, a royalty bidding system is in a second-best position after bonus bidding. The only costs of administration beyond the minimum found in the pure bonus-bid situation are: one, the requirements that production be measured, and two, the need to place a wellhead value on that production. Severe problems occur in the latter function where the oil or gas produced is not sold at the wellhead in an "arm's-length" transaction. This problem is compounded when there is no well-established market price in the relevant market area. This situation frequently occurs when the producer processes the oil or gas in his own refinery, or when products are traded rather than sold. For example, Alaska North Slope (ANS) oil is produced almost entirely by vertically integrated firms that either use this oil in their own refineries or trade it for oil in other locations. There is no ANS crude oil market at the wellhead. Further, the West Coast market is complicated by crude oil exchanges and by broad quality and locational differences making wellhead valuation difficult. In California, similar problems exist for the large Wilmington field near Long Beach, where oil produced is commonly utilized in refineries owned by the lessees.

If a sliding scale is used in royalty bidding, then administrative costs are likely to increase substantially. The lessor may determine that the lessee is constraining output in order to reduce his royalty obligation and the lessor may want to introduce a production monitoring program to enforce production at a level that maximizes current output, subject to physical recovery constraints. All of these problems lead to disputes and to litigation resulting in higher administrative costs for both lessee and lessor, resulting in a reduction or dissipation of the economic rent.

Relative to the objective of unambiguously identifying the high bidder, royalty bidding fails to pass the test. Royalty bidding payments to the government depend on the gross value of the oil produced. An efficient firm may either find less, or produce a lower percentage of oil (or gas) in place. Consequently, the government may receive a lower income flow from the inefficient firm that has submitted the highest royalty bid.

The high royalty bid may be meaningless when it is subject to renegotiating in the event that the lease is economically inoperable due to the royalty requirement. In both California and federal regulations, renegotiation is specifically allowed. Section 6827.2 of the California statute provides as follows:

> ... If, after the holding of a public hearing, the commission finds that continued production from a lease is in the best interest of the people of California and that such production is economically unfeasible under the terms set forth in the lease, the commission may renegotiate the lease to reduce the minimum royalty rate ... (PRC 6827.2).

Similar provisions governing federal leases render meaningless the eight high royalty bids received in the 1974 royalty bid sale, except as a means of allocating the leases between the competing bidders. Bids requiring royalty payments between 52% and 82% of wellhead value (gross income) are economic only in the event that a giant field is found. In the meantime, the royalty bid is merely a low-cost option, perhaps pending discovery of oil or gas on an adjacent lease. This appears to be the case on the 1974 royalty bid leases. Seven of the eight leases were returned to the government without producing oil or gas. One was unitized with an adjacent lease bearing a 16⅔% royalty and is producing gas from a common reservoir.

Pure profit-share bidding

Under pure profit-share bidding, the rent transfer function may or may not affect investment or output decisions, depending on how profit is defined. At least four choices are available:

First, profit may be defined in the conventional IRS sense, that is, in the same manner that taxes are calculated. In this event, investment decisions will be distorted because the profit-share payment is, in effect, an addition to income taxes beyond the approximately neutral effect of the latter. Investment decisions are made on an after-tax basis. Some leases that promise after-tax profits above the cut-off point will fall below that point and be abandoned due to the higher "tax" rate. One might argue that the oil or gas resource will still be there for later production. However, at a later point, the tract must be re-drilled and bear new marginal cost, whereas on the first occasion the drilling cost had become a sunk cost and would not affect the initial decision to produce or not produce.

Alternatively, profit subject to sharing may be defined in accordance with the "fixed-capital recovery system," the "investment account system,"[6] or the "annuity capital-recovery system."[7] In all three of these systems, the capital outlay, with or without an imputed interest charge, is subtracted from revenue before computing the profit-share to be paid to the government. In this event, new investments that might enhance recovery would not be incorrectly restricted. Instead, some otherwise uneconomic investment would be made. This will occur because these systems allow full recovery of the investment in productive leases, plus a return on the investment. Because the lessee is allowed full recovery of investments which fail to yield revenues in excess of the

---

[6] This system, authorized for Alaska by the Alaska Legislature in November 1979, is now in use.

[7] For a discussion of the alternative profit-share systems, see McDonald (1979), pp. 102-105.

investment outlay, the lessee is likely to become less cautious than he would be under bonus bidding where he bears the full cost (before taxes) of his mistakes. Whether economic investments are discouraged, or uneconomic investments are encouraged, the effect on economic rent is the same—the available economic rent is reduced.

The two primary advantages of profit-share bidding are that the front-end loading problem inherent in bonus bidding is avoided, and that, as a contingency payment system, a close correspondence would be established on an individual sale basis between the value of a lease and the payment to the lessor.

A potential, subtle advantage of profit-share bidding is that overzealous and uneconomic regulation might be restrained for the reason that such regulation reduces profit and the government shares in that loss. There appears to be a widespread misconception that regulations, particularly environmental regulations, can be imposed without regard to social costs and benefits. This misconception appears to be based on the incorrect conclusion that oil companies bear the economic cost. Some of the private cost, in the short run, may be borne by lessees. In the long run, it is highly probable that consumers will bear this private cost. The social cost in excess of the social benefit is a net loss to society as a whole. The imposition of regulations having social costs greater than benefits should be avoided, regardless of the leasing regime.

Profit-share bidding entails much larger administrative costs than encountered in the royalty system because, in addition to the need to determine the value of production revenue as in royalty payments, investments and costs must also be monitored and annual audits must be conducted. Unfortunately, no reliable estimates are available of the administrative costs borne by either the lessor or the lessee. In its proposed net profit-share leasing regulations under the "investment account rule," the Department of Energy (DOE) speculated that a lessee firm would incur accounting system modification costs estimated at between $50,000 and $150,000. In addition, annual administrative costs were estimated at between $25,000 and $30,000 per lease. Apparently, potential litigation was not considered by DOE. Also, DOE provided no estimate of increased administrative costs for the government. Firms responding to the proposed regulation countered that administrative costs were underestimated, but no estimates were given by the responding firms.

A recent evaluation of the Net Profit Share Leasing (NPSL) system practiced in the 1978–1983 years under the 1978 Outer Continental Shelf Lands Act provides some rough estimates of administrative costs incurred for both lessor and lessee. For the government, the Resource Consulting Group, Inc. (RCG) drew on Department of the Interior estimates that a single audit would require between one-half and 3 man-years of effort costing between $20,000 and $300,000 for labor alone on one tract (Resource Consulting Group, 1982, p. 4.4). This estimate does not include audit expenses other than labor, nor does it include any negotiation or litigation cost that would result from disputes between the parties. The reason for the uncertainty is that as of the date of the report "none of the tracts leased under the NPSL system have reached the production stage" (*Ibid.*, p. 4.1).

Interviews with lessees produced three cost estimates. A major firm estimated a cost of $30,000 annually per tract during the exploration period, plus $175,000 annually per tract during the development and production years. These estimates do not appear to include any costs for litigation. A second major firm estimated additional administrative and accounting cost of at least $27,500 per tract annually, plus $14,000 per tract for conducting inventories and assembling additional information for joint interest audits (*Ibid.*, p. 3.39). A third firm noted that the estimate of $30,000 annually per tract for accounting and administrative expenses was too low (*Ibid.*).

Apprehension that having the government as a partner with adversarial interests would lead to expensive litigation caused most firms to oppose the NPSL system (*Ibid.*, p. vi). This opposition led four out of 14 firms to decide not to bid on NPSL tracts (*Ibid.*, p. 3.3).

The source of the heavy administrative costs is what economists call the "agency problem." Where there is a divergence between the interest of the lessor and the lessee as an agent, one must expect that the agent will behave in a manner that serves his own interest first. The agency problem leads to monitoring costs by the lessor in an effort to prevent the lessee from pursuing such opportunistic behavior. The divergence of interest arises out of the fact that the cost of some projects is deductible from revenues, and, where a lease is profitable, the costs are shared according to the profit-share percentage. However, the benefits of the project may accrue primarily or entirely to the lessee firm. An experimental exploration or production system is an example.

The agency problem is compounded by income taxes and any severance taxes and royalties. Let us take a simple example involving only corporate income taxes at a 34% rate and a 50% profit-share. The incentive to operate efficiently is a reward of 66 cents per dollar of profit after income taxes. Paired with a 50% profit-share requirement, $1 of profit reward generated by an efficiency effort is reduced to a reward of 33 cents after taxes and profit-share. The other half of the coin would state that government bears 67% of the cost of wasteful activity. This is a special case of the externality problem. Where benefits accrue primarily to the lessee, but the costs are shared between governmental and the lessee, some investments and expenditures will be made because the benefits to the lessee exceed the cost-share borne by the lessee. The nation learned this lesson during the Korean War when an excess profits tax was levied on corporate income with the result that the highest tax bracket became 82%. Under such conditions, leasing a corporate jet, business conferences in Hawaii, and similar "perks" that benefitted management, were commonplace for the reason that they cost only 18 cents on the dollar after taxes.

We will provide illustrations of these potentially wasteful expenditures below. However, it should be understood that the list is not exhaustive.

(1) When a company wishes to do some research and development (concerned with oil exploration and production) where offsetting benefits are assigned a low probability of success, it is more likely to do so on leases involving profit-share payments, where the losses are shared with the government. If the experimental work is successful, it can then be adopted on other properties controlled by the lessee (thereby producing benefits for the lessee).

(2) Where a company has a mixture of high-quality and low-quality drill rigs or drill ships, it is likely to use the poor equipment on the profit-share lease and reserve the best equipment for other company operations.

(3) Where a company needs to train crews in drilling and reservoir development, it is likely to do its training on profit-share leases.

(4) "Gold-plating" is likely to occur on profit-share leases, where the share paid to government is high and the retained share is low. Evidence of this practice may be found in the Long Beach (Wilmington) field where profit-shares paid to the government are extremely high. In this instance, attractive palm tree-studded islands have been created to shield the production from public view, in lieu of the usual and frequently unsightly production equipment, towers, and piping. Substantial public relations benefits flow to the operating firms, but the expense is borne almost entirely by the City of Long Beach.

(5) In the event of oil equipment supply shortages, such as occurred in 1973 and 1974, profit-share lessees are likely to allocate available supplies to their non-profit-share leases first. The opportunities to shift expenditures to profit-share leases are limited only by the imagination of the lessee and expensive policing of expenditures by the lessor.

Some of the shifts suggested above are merely the equivalent of income transfers and therefore involve no social costs to society. However, any shift which leads to inefficient investments or operations creates social costs. In any case, the economic rent available to the lessor is reduced by shifting of the type illustrated above. Such shifting forces the lessor to protect his interest against agency behavior by monitoring investment and operating decisions. This leads to monitoring costs. Administrative interference in the operation of a profit-share agreement becomes a necessity. The Long Beach–Wilmington contract provides as follows:

> . . . The City Manager . . . shall exercise supervision and control of all day-to-day unit operations . . . and . . . shall make determinations and grant approvals as he may deem appropriate for the supervision and direction of day-to-day operations of the Field Contractor, and the Field Contractor shall be bound by and shall perform in accordance with such determination . . . (Contractor's Agreement, Article 4).

The Federal government has recognized that the divergent interest problem leads to increased monitoring. In the Energy Department's "Final Rule-making" regulations, the point was made that: in a very real sense, the USGS (U.S. Geological Survey) represents the interest of the public in the exploration and development of its resources and already has an active role in OCS decision. As future sharer in net profits, USGS must ensure that any sharing arrangement make provisions for monitoring expenses incurred during exploration, development, and production, in order to ensure an equitable division of net profits (Federal Register, 1980, p. 36790).

The regulations require approval of all expenditures subject to fixed capital recovery. The federal system allows full recovery of these capital expenditures, plus a specified rate of return for the operator, providing offsetting revenues are subsequently generated from the lease. In the case of subsequent operating expenditures that are fully deducted against income but without an additional payment for the use of capital, the regulations state that "The lessee retains absolute discretion to incur such additional costs as appear warranted." However, all expenditures are subject to audit within 3 years following the expenditure. In instances where the government attempts to deny prior outlays, legal challenge must be expected. Such litigation will impose heavy costs on both the lessee and the government, and serve to reduce or dissipate economic rents.

In the case of wages and salaries charged to the profit-share lease, the regulations state vaguely that "employees need be engaged in (profit-share lease) operations continually only for a specified period of time, such as a month, week, or pay period. These wages and salaries as well as other enumerated personnel costs are then includable" (*Ibid.*, p. 36794). The lessee will have a clear incentive to charge wages and salaries to the profit-share lease, to the limit of credibility. The lessor has a similar clear incentive to police all expenditures. One must expect costly monitoring and litigation to follow.

In performing the essential functions of post-lease exploration, drilling, development, well reworking, and the like, some of the cost may not be specific to particular leases, (transportation, for example). This leads to the problem of allocating costs between leases. The usual incentives arising out of the agency problem will lead to litigation.

The lessee normally owns equipment that will be used on profit-share leases. In the absence of an "arm's-length" rental transaction, problems will arise as the lessee attempts to maximize his rental charge on the profit-share lease. The regulations allow equipment rents to be based upon "actual costs of acquisition, construction, and operation . . . subject to the ceiling of average commercial rates for similar equipment and facilities prevailing in the vicinity . . ." (*Ibid.*, p. 36792). The ceiling will be difficult to apply due to differences in location, environmental hazards, state of the market, type of equipment, etc. The divergent interests of the parties will lead to litigation.

Finally, there will be litigation over what litigation costs may be charged against the profit-share lease. Where the lessee uses in-house legal counsel, allocation of such joint costs becomes highly subjective. Regulations must fall back on pre-litigation wording—joint costs must be "reasonable and equitable," but these words will have to be interpreted by the courts.

In sum, the agency problem leads to detailed regulations. The lessor must specify operating behavior as well as the accounting of all costs in order to protect lessor interests. This necessity leads to litigation as well as administrative costs to be borne by both the lessee and the lessor. In the first event, economic rents, as illustrated in Figure 7-2, are reduced. In the second, economic rents are dissipated by the government in ways that do not benefit the public. There are few or no comparable problems of this sort in the cases of bonus payments.

Simulation models have been constructed to estimate the economic rent that would be collected under profit-share bidding relative to the conventional bonus bidding with a fixed royalty. Reece concluded that the government would capture substantially greater rent under profit-share bidding (Reece, 1979).[8] However, he assumed that "the total rent recovered by society would be the same under alterative leasing systems." In effect, Reece assumed that the lessee's exploration and development program was *independent* of whether royalty, profit-share, or bonus bidding was used. Furthermore, he ignored the problems associated with *defining* a competitive equilibrium under the pure royalty and pure profit-share bidding systems.

Finally, Reece abstracted from the *sequential* attributes of exploration by assuming that the lessee would incur only a one-time *fixed costs* in evaluating the lease after the auction, exploring the lease, and developing the lease. However, because of the no-cost option attributes of pure royalty and pure profit-share bidding, the lessee may exercise his right to abandon the lease without penalty at any time. Clearly, if firms differ in their efficiency attributes, the exploration and production profile of the lease will depend on who is initially assigned the right to explore and develop the lease.[9] These assumptions are sufficient to tilt the conclusions in favor of profit-share bidding. The critical issue in such models is the validity of the assumptions. Our evidence indicates that administrative costs for profit-share operations are both high and wasteful.

The high bidder ambiguity problem identified for royalty bidding also applies to profit-share bids. In the example in Table 7-6, Firm A, bidding a 50% profit-share, will be the winner over Firm B bidding a 40% profit-share. However, Firm A is less efficient (produces less revenue at higher cost) and generates less profit. Thus the payment to the government by the apparent high bidder is less than the second-highest bidder would have paid. If the government is to have confidence that by accepting the apparent highest bid it will

---

[8] For similar models and assumptions, see State of Alaska, Department of Natural Resources (1979); and Kalter, Tyner, and Hughes (1975).

[9] Pickett provides a discussion of the impact of alternative leasing regimes on the lessee's exploration decisions in the context of an option valuation model (Pickett 1983).

receive the largest payment, it must conduct a separate analysis in which it evaluates the efficiency attributes of the bidding firms. As a practical matter, this is impossible. Consequently, the high bidding firm in royalty and profit-share systems will get the lease but will not necessarily produce the highest payment to the government.

Work-program bidding

Work-program bidding is a system wherein the bidder is invited to state exactly what he would do if awarded the lease. This system was authorized for use by the federal government in 1978 legislation that was allowed to expire in 1983. Work-program bidding is not currently permitted by California statute. Where allowed, government may suggest areas in which the bidder is required to specify his proposed work-program. For example, proposals may be requested concerning the number of wells that would be drilled; whether new onshore investments would be made in separation plants, refineries, petrochemical plants, etc.; whether sub-sea completions would be used; and the like.

In Alaska, bidding on a proposed work-program regime requires a statement of the dollar amount for the program. Presumably this means that the bidder would cost-out his entire exploration, development and production program. A winning lessee would then be required to post a bond to ensure that the commitment will be faithfully discharged.

Australia uses work-program bidding for leasing its submerged lands under authorization of its Submerged Lands Act of 1967. Legislation there allows exploration permits covering a maximum of 400 blocks for each permittee (about 31,500 square kilometers or 7,780,500 acres). This is an extremely large area, 1,350 times as large as the California and U.S. government maximum lease size (5,760 acres). The initial term of the exploration permit is 6 years. At the end of that period, half of the area must be relinquished. Extensions are granted for subsequent 5-year periods. At the end of each 5-year period, half of the remaining land area must be relinquished.

TABLE 7-6   Hypothetical Example of a Profit-Share High Bid That Fails to Maximize Revenue to the Government.

	Profit-share Bid	Efficiency Test Revenue – Cost = Profit	Payment to Government
Firm A	50	100 – 60 = 40	20
Firm B	40	115 – 55 = 60	24

Exploration permits in the important Western Australia OCS area are administered by the adjacent state government. Leases there are awarded to the high bidder on the basis of proposed exploration programs. The criteria for selecting the high bidder are:

> . . . the amount of seismic surveying to be undertaken and the number of wells to be drilled on a yearly basis. Other criteria include the timing of the proposed exploration (e.g., work done early in the program is given a higher weighting than later work), the extent to which the minimum work-program reflects the available technical information in the area, the extent to which the program seeks to assess new prospects in previously unexplored parts of the area, and the adequacy of the financial resources and technical expertise available to each applicant (Government of Western Australia, 1984, p. 18).

The government acknowledges that

> . . . during the course of the 5- or 6-year period of the permit, some modifications of the work-program may become necessary for operational reasons. Rescheduling or variation of the yearly work and expenditure commitment will be allowed at the discretion of and only with the prior approval of the relevant authority (*Ibid.*).

With prior authorization for revision (downward only?) of the work-program bid, bidders have an incentive to promise extremely generous exploration programs in order to increase the probability that they will be awarded the permit.

Even without prior knowledge that high bids will be revised downward, bidders would have an incentive to promise exploration programs much more costly than could be justified on the basis of probable benefits and costs. Australia imposes royalty and "excise" tax requirements similar to those in the United States. Prospective bidders would estimate the present value of the probable total revenue, subtract the present value of all necessary costs including required royalty and tax payments, leaving a residual economic rent. In a bonus-bid system, this amount would be approximated in the high bid. But under work-program bidding incentives, bidders would offer an incremental work-program having a cost equal to or less than the amount of the residual economic rent as defined above. Thus, the work-program bidding system substitutes exploratory expenditures for what would otherwise be a residual payment to the state. The state, in lieu of this cash flow, receives an exploratory program having costs in excess of benefits. The only exception to this result would occur if the residual economic rent was estimated at or near zero.

The disadvantages of this system for economic rent maximization are overwhelming. The optimum level of investments in well drilling, onshore facilities, field development, and in other areas cannot be specified in advance of knowledge about drilling results. If initial drilling is unsuccessful, subsequent drilling should normally be scaled down, relative to highly successful initial drilling. Similarly, optimal field development investment plans can be determined only as the extent, quality, and value of reserves become known.

Work-program bidding is based on the assumption that the leasing bureaucracy is in a position to know the socially optimal line of development.[10] However, this group is generally neither trained nor experienced in making investment decisions, either from a private or a social perspective (there may be no difference). Yet it must decide which applicant is to be awarded the lease. Experience with this system in the North Sea fields has shown that bidders are forced to determine what kind of work plan is favored by the decision-making bureaucracy and then bid generously in that (or those) favored area(s) (Dam, 1976).

To the extent that work-program bidding results in wasteful investments or operating costs, dollar-for-dollar reductions (in present-value terms) occur in cash-bonus or other cash payments to the government. In terms of Figure 7-2, the work-program bidding system causes unnecessary expenses to rise to the point where economic rent (not captured by a fixed royalty requirement or the excise tax payment) becomes zero. Thus, work-program bidding fails the test of maximizing and collecting the full economic rent.

Work-program bidding, like profit-share bidding, imposes heavy administrative requirements on both lessor and lessee. The program proposed by each bidder must be carefully defined in his bid statement. It must then be carefully evaluated by the lessor in order to identify the "high bidder." Following designation of the winning bid, the lessor must monitor the lessee to ensure that he performs in a manner that is not less than his bid statement guarantees. Performance of this duty requires that the work-program be specified in great detail in order to minimize disputes as to the promised work-program. The same agency problems exist here as identified above relative to profit-share bidding. After winning the lease, the lessee has an interest in minimizing his costs for

---

[10] That is, the bureaucracy is a substitute for the *sequence* of competitive secondary markets for OCS leases. For a discussion of secondary markets under alternative leasing regimes, see Pickett (1983).

projects that may be of importance to the bureaucracy, but of little importance to the lessee. One must expect that monitoring the work-program will lead to heavy administrative costs, including expensive litigation.

Except in the odd situation where all elements save one in the work-program bid by two or more firms are identical, there is no way to objectively determine the high bidder. Judgment must be subjective. Where the winning bid in a work-program bid is subject to later renegotiation in the event that the lease can be shown to be uneconomic under the specified work-program, the high bid will normally offer a work-program that is uneconomically expensive in order to win the lease. Thus, the bid does not correspond with the work later performed and the bid neither identifies unambiguously the high bidder nor the most efficient operator.

Threshold rate bidding in a Resource Rent Tax system

This bidding system is considered only as an element of a fixed-payment system known as a Resource Rent Tax (RRT). The RRT system will be discussed in the section below.

The threshold rate bid is merely a device to identify the winning bidder in the RRT system. The threshold rate is the minimum acceptable rate of return that a firm would earn as the operator of a lease in the RRT system. The firm offering the lowest threshold rate would be declared the winner.

Alternatively, identification of the winning firm might use a cash-bonus bid mechanism. Most of the economic rent would be captured in the RRT. Consequently, the amounts bid under the bonus alterative would be relatively small.

There are no special problems associated with either bidding system apart from the RRT itself. The significant economic problems associated with the RRT system will be pointed out in the following section.

Summary

From the perspective of maximizing and collecting the economic rents, a cash bonus appears to be the superior bid variable, and profit-share bidding a distant second-best. Royalty bidding and work-program bidding should be rejected without further experimentation on the argument that these systems substantially reduce the economic rents available to society.

The high bidder is unambiguously indicated in bonus bidding, and imperfectly identified in royalty and profit-share bidding. Only in bonus bidding is there a theoretical basis for asserting that, in the long run, high bidders are the most efficient operators. In work-program bidding, ordinarily no objective test of high bid is available. The incentive of bidders is to promise more of some exploration, development, or production variable than would be economically efficient. This same fault applies to royalty bidding if rate renegotiation is allowed. Some economic waste arising out of work-program bidding is reduced (or even eliminated) when the nominal high bid is scaled down in post-bid negotiations between lessee and lessor.

From the viewpoint of minimizing administrative costs, bonus bidding is preferable, with royalty bidding in second position. Profit-share and work-program bidding involve high and wasteful administrative costs for both lessor and lessee and reduce or dissipate economic rent. No reliable estimates are available indicating the amount of such costs.

A primary merit of royalty and profit-share bidding is that if no oil or gas is found, no payment is made to the government. When commercial quantities of hydrocarbons are found, then payments to the government are correspondingly large. However, bonus bidding is superior to royalty bidding in this respect because in bonus bidding, payments are based on estimated future revenues minus costs, whereas in the royalty case, payments are based on gross income only, leading to premature abandonment, abandonment of productive reservoirs, and rejection of economically justified investments. Bonus bidding fails in this test, on a tract-by-tract basis. For example, Exxon, Mobil and Champlin bid and won the Destin Anticline sale offshore west of Florida in 1973 with a high bid amounting to $212 million, and never found a commercial oil or gas deposit. In contrast, the Prudhoe Bay field, containing the largest oil reserve ever found on the North American continent, was obtained at a bonus cost of only $10 million for its estimated 9.6 billion barrels of oil and 26 trillion cubic feet of natural gas. Again, in contrast, this bargain win was followed by enthusiastic bidding for North Slope leases in 1969 where bidders paid over $900 million for leases without even one announced commercial discovery.

While bonus bidding clearly fails to produce cash-bonus bids approximating real values on a tract-by-tract basis, as we have demonstrated earlier, *in the aggregate* there is a close correspondence. Thus, diversification available to the State of California and the federal government (i.e., selling many leases to many buyers) yields the same benefits as if tract-by-tract correspondence existed.

Similarly, diversification on a firm-by-firm basis produced approximately normal returns to the firms. On the basis of the 1,223 leases issued from 1954 through 1969 in the Gulf of Mexico, all of the big-20 firms had positive rates of return on their leases. The after-tax returns ranged from 3.37% to 20.88% and the number of leases or lease interests per company ranged from 13 to 167 (Mead, Moseidjord, Muraoka, and Sorensen, 1985, p. 82). From an economic perspective, an evaluation of the national interest indicates that the lack of correspondence on a tract-by-tract basis is of negligible importance. However, politicians appear to be embarrassed by the occasional Prudhoe Bay situation, where the lessee obtains a bargain even though the larger record shows that approximately 62% of the tracts leased were dry and worthless, while another 16% were productive but worthless.

In the case of work-program bidding, little can be said about the correspondence issue. The terms of payment depend on the criteria of the decision-making bureaucracy and the bidders, as well as on the fixed payments.

ALTERNATIVE FIXED PAYMENT REQUIREMENTS

In the foregoing, we have discussed five bid variables. Leasing regimes commonly pair a bid variable with one or more fixed payments. Most commonly, a cash-bonus bid variable has also required a fixed rental payment which must be paid annually until production begins. The rental requirement is then replaced by a royalty payment as soon as production begins.

The first purpose of the bid variable is to determine the winning bidder. Its second purpose is to specify part of the payment flow to the government. But the latter is determined jointly with any fixed-payment requirements. The terms of the fixed payments are specified in advance of bidding, whereas the bid variable is determined in the auction process.

The economic analysis given above applies equally to the selection of fixed payments. The important difference is that the amount of the fixed share, being controlled by the government, can be set in a manner that avoids the problems of very high bids for bonus, royalty or profit-shares. The use of fixed royalty or profit-share payments leads to reductions in economic rents and should be avoided because of the distortions they cause in the form of revenue misallocation and avoidable administrative expenses.

Governments, both federal and state, commonly specify work-program requirements in the lease contract. They should also be avoided unless they produce social benefits in excess of social costs. If this rule is violated, there will be a reduction in social welfare. In this event, the requirements are wasteful of resources and eventually cause reductions in living standards. Whether wasteful or useful, work-program requirements involve costs. If the money costs of the work-program requirements are known in advance of bidding, the full present value of such costs will be subtracted from the bid amount in the case of a bonus bid. Thus, government bears the full money cost of its work-program requirements known in advance of bidding.

In formulating public policy relative to bidding regimes, it should also be understood that other tradeoffs exist between the specified bid variable and the fixed payments. If the traditional 16⅔% royalty is eliminated, competition will force firms to bid higher bonuses. Because economic inefficiency occurs as a result of royalty payments, its elimination would cause the present value of bonus bids to increase more than the loss in present value of royalty payments. Similarly, if a zero profit-share is replaced by some positive profit-share payment requirement, the present value of the bonus bid will be reduced by more than the present value of the profit-share payment. This result follows from the point that profit-share requirements impose avoidable administrative costs on the lessee.

In Australia, where an economically inefficient work-program bidding system has been used with various tax and royalty fixed payments, there is experimentation with a new system utilizing a Resource Rent Tax (RRT), sometimes called a Resource Rent Royalty (RRR).

A pure RRT system would collect a tax equal to a specified tax rate multiplied by the net cash flow from each property, after a "threshold rate of return" to the operator has been added to other costs. Where a productive reservoir is discovered and produced, the lessee would be allowed a full recovery of all his outlays (no distinction between capital investment and annual expenses) plus a threshold rate of return on his accumulated net cash outlays, before the RRT is levied. The threshold rate would be computed on the net cash invested at the end of each year and added to other net costs. As long as a net cost is carried forward, the RRT would be zero. To compensate for a series of dry holes, a high threshold rate of return would be expected.[11]

The idea of an RRT has stimulated interest among economists because it promises tax neutrality; that is, it has no resource misallocation effects. Payments are not "up front" and therefore the RRT has no tendency to restrict competition for leases. It shifts risk from the lessee to the government as the resource owner. It solves the problem that Leland (1978) alleged, that of risk-averse behavior by private firms leading to high discount rates and consequent low bonus bid values.

Under a pure RRT, allocation of leases among two or more interested lessees would presumably be done by a bureaucracy. However, this problem may be solved by "impurely" setting a high RRT rate which leaves some positive present value, then allocating leases on the basis of a bonus bidding system. This procedure has been proposed in at least three papers (Garnaut and Ross, 1977; Emerson and Lloyd, 1983; and Emerson and Garnaut, 1984). Finally, as suggested in the section above, allocation might be based on a threshold rate bid, with the lease being awarded to the lowest bidder.

However, the RRT system leads to a reduction in economic rent because, like a profit-share system, it requires high administrative and legal costs for the lessee. It also leads to rent dissipation, because the government must incur heavy supervision and legal costs. These lessor and lessee costs are nil in the pure bonus-bid case.

Again, like a profit-share system, investment and expense outlays must be monitored to assure that such outlays, while of value to the lessee, are not expense padding from the lessor perspective. The agency problem outlined relative to profit-share bidding applies equally to the RRT system. Where the operator is an integrated firm, stated revenues must be examined because they will not be "arm's-length" prices.

Finally, the RRT system differs from the public utility rate-making problem in only one important respect. In the latter, the objective of regulation is to mandate a system of product prices such that the fair rate of return will be earned on the rate base. In the RRT case, the objective is to determine the amount that must be paid to the government such that when all other costs are subtracted from revenues, the lessee will earn only a fair rate of return on his portfolio of lease investments.

---

[11] For a discussion of the arguments in favor of an RRT, see Garnaut and Ross (1982), pp. 166–167.

CONCLUSIONS AND POLICY RECOMMENDATIONS

1. On the basis of resource allocative efficiency, the optimum bidding and leasing system appears to be a *pure* cash-bonus bidding regime (i.e., without royalty or other fixed payments). It would maximize the economic rent and minimize the government's administrative costs thereby minimizing the dissipation of its economic rent.

The main negative feature of a pure bonus-bidding system is that, on a tract-by-tract basis, payments to the government do not correspond precisely with the economic rent. However, due to the fact that the government sells many leases, returns to the government correspond closely with the amount of economic rent that is embodied in the resource.

To the extent that a fixed payment must be added to a bonus-bidding system, an annual rental payment per acre would involve only minor additional social costs. Fixed royalty payments should be deleted from the lease sale regime because such payments increase administration costs and reduce oil and gas production that would otherwise produce gains for society in excess of social costs. Nor should a fixed profit-share or a resource rent tax be added. Large social costs occur due to heavy administrative costs for both lessee and lessor.

2. As a second-best system, we recommend the conventional federal system of bonus bidding with a fixed royalty. The federal system for leasing OCS oil and gas resources has been in use since 1954, is reasonably efficient, and has produced over $93 billion of bonus, rent and royalty income for the U.S. treasury. Bonus payments alone amount to 53% of the total, or about $55 billion.

Both the federal and state governments could improve performance under this conventional system by taking two steps—first, by reducing the fixed royalty rate from $16\frac{2}{3}\%$ to $12\frac{1}{2}\%$ as provided in legislation, the abandonment before production problem would be mitigated; second, by exercising its legal power to reduce the fixed royalty rates ultimately to zero as exhaustion of the well, field, or tract approaches, the premature abandonment problem would be eliminated.

3. The third-best system would be bonus bidding with a fixed profit-share requirement. The principal social cost, causing a reduction in economic rents and their dissipation, is the high and wasteful administrative cost of collecting the profit-share.

4. As a poor fourth-best system, profit-share bidding would have the merits of a close correspondence between payments to the government and the value of the lease to the lessee, and shifting risk to the government via this contingency payment. However, high profit-share payments involve heavy administrative costs; and the higher the percentage, the more litigation is likely to occur due to disputes between lessor and lessee. Disputes occur relative to oil and gas prices and therefore revenue, where lessees are vertically integrated, and in allocating investment and expenses between parent and subsidiary firms, and between profit-share and non-profit-share leases. (For a more favorable view of profit-share bidding and fixed payments, see McDonald 1984; Gaffney 1977; Leland, Norgaard, and Pearson 1974.)

Systems with little or no merit that should be avoided entirely include royalty bidding and work-program bidding. Both impose heavy social costs on the nation.

5. Rules such as the 5-year production rule in federal OCS leasing and the 3-year drilling rule in California should be eliminated and avoided. The lessee should be free to decide when to produce from his lease. If the lessee is a profit-maximizing operator, he will maximize the present value of his leasehold. This is precisely the economic definition of resource conservation. A rule that forces early development of petroleum resources may not be consistent with a conservation goal.

6. Government regulations to protect the environment or serve other objectives should be justified with evidence that such regulations create social benefits greater than social costs. Any regulations which have the opposite effect do not serve the conservation goal. The economic cost of the latter is paid by the citizens of the nation, not by lessees.

7. Use of leasing policies to increase competition in the lease sale market do not appear to be necessary. Empirical studies indicate that competition is effective in the OCS auction markets under conventional leasing conditions.

8. Uncertainty and ambiguity in lease contracts should be avoided because uncertainty leads to higher discount rates, in turn leading to lower present values for leases. Thus, the government receives lower bonus bids under conditions of increased uncertainty.

REFERENCES

Contractor's Agreement, Long Beach Unit, Wilmington Oil Field, California.

Dam, K. W. (1976). *Oil Resources*. Chicago: University of Chicago Press.

Emerson, C. and R. Garnaut (1984). "Mineral Leasing Policy: Competitive Bidding and the Resource Rent Tax Given Various Responses to Risk." *The Economic Record*, Vol. 60, No. 169.

Emerson, C. and P. J. Lloyd (1983). "Improving Mineral Taxation Policy in Australia." Economic Record, Vol. 59, No. 166, pp. 232–244.

Federal Register. 1980. Vol. 45, No. 106, p. 36790.

Gaffney, M. (1977). "Oil and Gas Leasing Policy: Alternatives for Alaska in 1977." Unpublished report to the State of Alaska, Department of Natural Resources. February.

Garnaut, R. and A. C. Ross (1975). "Uncertainty, Risk Aversion and the Taxing of Natural Resource Projects." *The Economic Journal*, June.

Garnaut, R. and A. C. Ross (1982). *Taxation of Mineral Rents*. Oxford: Clarenden Press.

Government of Western Australia, Department of Mines, Petroleum Division (1984). "Petroleum in Western Australia." January.

Kalter, R. J., W. E. Tyner, and D. W. Hughes (1975). "Alternative Energy Leasing Strategies and Schedules for the Outer Continental Shelf." Research Paper AE 75-33, December. Ithaca, New York: Cornell University.

Leland, H. E. (1978). "Optimal Risk Sharing and the Leasing of Natural Resources with Application to Oil and Gas Leasing on the OCS." *Quarterly Journal of Economics*, Vol. 92.

Leland, H. E., R. B. Norgaard, and S. R. Pearson (1974). "An Economic Analysis of Alternative Outer Continental Shelf Petroleum Leasing Policies." Unpublished report to the Office of Energy R&D Policy, National Science Foundation. September.

McDonald, S. L. (1979). *The Leasing of Federal Lands for Fossil Fuels Production*. Baltimore: Johns Hopkins University Press for Resources for the Future.

Mead, W. J. and A. Moseidjord (1984). "Competitive Bidding Under Asymmetrical Information." *Review of Economics and Statistics*, Vol. LXVI, No. 3.

Mead, W. J., A. Moseidjord, D. D. Muraoka, and P. E. Sorensen (1980). "Additional Studies of Competition and Performance in OCS Oil and Gas Leasing of the U.S. Outer Continental Shelf." Unpublished final report to the U.S. Geological Survey, Contract No. 14-08-0001-18678.

Mead, W. J., A. Moseidjord, D. D. Muraoka, and P. E. Sorensen (1985). "Offshore Lands, San Francisco." Pacific Institute for Public Policy Research.

Mead, W. J., A. Moseidjord, and P. E. Sorensen (1983). "The Rate of Return Earned by Lessees under Cash Bonus Bidding for OCS Oil and Gas Leases." *The Energy Journal,* Vol. 4.

Mead, W. J. and D. D. Muraoka (1987). "Diligence Requirements in Federal Natural Resource Sale and Leasing." *Natural Resources Journal,* Vol. 27.

Mead, W. J. and G. C. Pickett (1984). "Federal Leasing Policy." In S.F. Singer, ed., *Free Market Energy.* New York: Universe Books.

Pickett, G. C. (1983). "An Option Valuation Model of Bonus Bidding and Profit Share Bidding for Offshore Oil and Gas Leases." Ph.D. dissertation. Santa Barbara: University of California.

Reece, D. K. (1979). *Leasing Offshore Oil: An Analysis of Alternative Information and Bidding Systems.* New York: Garland Publishing.

Resource Consulting Group, Inc. (1982). "Issues Associated with the Use of the Net Profit Share System for Leasing Outer Continental Shelf Oil and Gas Acreage." Prepared for the Department of the Interior, September 27.

State of Alaska, Department of Natural Resources, Divison of Oil and Gas (1979). "A Report to the State of Alaska."

U.S. Department of Energy, Energy Information Administration (1991a). *Monthly Energy Review.* March. DOE/EIA-0035(93-1). Washington, D.C.: Government Printing Office.

U.S. Department of Energy, Energy Information Administration (1991b). *Annual Energy Outlook.* DOE/EIA-0383(91). Washington, D.C.: Government Printing Office.

U.S. Department of the Interior, Minerals Management Service (1990). "Federal Offshore Statistics: 1989." MMS 90-0072. Washington, D.C.: Government Printing Office.

U.S. Department of the Interior, Minerals Management Service (1984). "Outer Continental Shelf Lease Sales and Evaluation of Alternative Bidding Systems." Annual Report to Congress, Fiscal Year 1983. Washington, D.C.: Government Printing Office.

U.S. House of Representatives (1977). "Report of the House Ad Hoc Select Committee on the Outer Continental Shelf." H.R. Report No. 95-590. Washington, D.C.: Government Printing Office.

Wilcox, S. N. (1975). "Joint Venture Bidding and Entry into the Market for Off-Shore Petroleum Leases." Ph.D. dissertation. Santa Barbara: University of California.

Wilson, R. B. (1978). "Management and Financing of Exploration for Offshore Oil and Gas." *Public Policy,* Vol. 26, No. 4.

THE PERFORMANCE OF
8 THE DISTRIBUTION
SYSTEM FOR GASOLINE

William S. Comanor

INTRODUCTION

The search for a more secure and regular supply of energy invariably focuses on gasoline as the predominant fuel in the transportation sector. This single product accounts for a major share of energy demand and is therefore a critical factor in our concerns over the supply of energy.

In this chapter, we explore the structure and performance of the distribution system for gasoline. Sources of energy, whether new or old, contribute to the economy largely when received by final users. Furthermore, all distribution systems provide various services that are bundled with the product and supplied directly to consumers. The nature of these services and the implicit prices charged for them can have a substantial impact on consumer demand for both established and new products. For example, the breadth of distribution for an alternative fuel used to power automobiles could have a major impact on its ultimate acceptance in the market place. As a result, the distribution system is an essential part of both the questions raised and the possible solutions to our "energy problem."

The distribution of transportation fuels to common carriers such as airlines and railroads raises few issues. There are relatively few buyers in the market

William S. Comanor is with the Department of Economics, University of California, Santa Barbara, California.

place and fuels can be supplied at a number of fixed locations. On the other hand, the distribution of transportation fuels to trucking companies and private automobile users raises a host of issues. Fuel tanks are relatively small so there must be a vast number of distribution outlets in all locations. Moreover, gasoline is costly to store in large amounts, so distribution points must be resupplied frequently. Overall, what is required is a widespread and sophisticated distribution system that has its own special characteristics. For these reasons, and unlike many other commodities, gasoline is distributed jointly with only a few other products in a unique set of distribution outlets. These outlets, of course, are called service stations.

Because of this individualized system of distribution, the relevant costs may be relatively high. While distribution costs can be kept to 1 or 2% of the selling price for some consumer products, that is not the case for gasoline. For full-service gasoline distribution, by the end of the 1980s, these costs represented nearly one-third of the total value of the bundled product supplied to consumers, while in the case of self-service distribution, costs were lower but still accounted for nearly 10% of the consumer's total cost.

Of particular concern, therefore, is the distribution margin for gasoline, which is the difference between the final consumer price for the product and the delivered wholesale price plus any taxes that are levied. This margin must cover the costs of distributing the product, including both labor and capital costs as well as any profits earned at the distribution stage. In this chapter, we examine the factors that determine distribution margins for gasoline. As reported below, these margins have varied greatly throughout the 1980s and have had a major impact on consumer demands.

How distribution margins, including profits earned by station owners, relate to the wholesale price of gasoline is a relevant issue. Some critics have suggested that margins increase when gasoline is in short supply, which contributes to the higher prices paid by consumers. This charge concerns the efficiency of gasoline distribution and also the relationship between service stations and their suppliers. In the discussion below, we explore available evidence on this matter.

The focus here raises various questions for public policy. If the performance of the system is good and markets respond as expected in a competitive system, there should be little need for government intervention into the distribution sector. On the other hand, if there are factors that limit the performance of this sector, then government intervention could be beneficial. Actions could be taken to limit price increases, to fix or restrict the number of stations, or to alter existing ties between service stations and the major oil companies. Before reaching these issues, however, we first evaluate performance.

THE STRUCTURE OF DISTRIBUTION

A striking feature of the distribution of gasoline is the relatively minor degree of vertical integration in terms of direct ownership. Only 17% of total sales by refiners in the United States are made to final users, either through direct sales or refiner-owned retail outlets (U.S. Department of Energy, 1981, p. 113). The major oil companies have found it more costly to operate their own service stations and have generally relied on independent operators.[1]

While most service stations are independently owned, they typically have contractual ties with a major oil company who serves as their supplier. For the most part, current relationships are exclusive so that a particular station sells only a single brand of gasoline. In this case, new entrants at the wholesale stage must create their own resale network and cannot realistically rely on existing stations to distribute their product. However, distribution margins vary across stations, so there is no automatic control exercised by the major oil companies on consumer prices.

In contrast, independent marketers are more likely to own their own stations and exercise greater control over their final prices. While these firms generally set lower prices than the major oil companies, this is not always the case. In California, these sellers have retained a relatively constant share of total sales of between 18 and 20% throughout the 1980s.[2]

The most dramatic recent change in gasoline distribution in California, and indeed throughout the nation, is the sharp decline in the number of stations. From more than 22,000 in 1972, the number of station permits in California fell to 15,789 in 1980 and 11,776 in 1989 (California State Board of Equalization). The number of stations thereby declined by half in these years. This decline,

---

[1] For an earlier discussion of the relative costs of independent and company owned service stations, see Miller (1963).

[2] The aggregate share of U.S. gasoline consumption gained by independent marketers increased throughout the 1970s and early 1980s, and reached more than 35% in 1986. See the Lundberg Surveys.

however, was not associated with a fall in the quantity of gasoline sold. On the contrary, as the data presented in Table 8-1 demonstrate, gasoline sales in California increased from approximately 11 million gallons in 1980 to 13 million gallons in 1988, even though service station permits fell from over 15,000 to 12,000 in the same time period. In many instances, a sharp decline in the number of retail outlets follows from a decline in the demand for the product, but this has not been the case for gasoline sales in California.

Taken together, these data show that the average station sold more gallons of gasoline. To explore this matter further, we computed average gasoline sales per station, which are reported in Table 8-2. There has been a continued increase throughout the 1980s in the average number of gallons sold per station, and there is even some indication of increased dollar sales per station, although that result is confounded by the declining price of gasoline in the period. Thus, gasoline distribution has been carried on at slightly higher levels than before but with many fewer outlets.

This change has taken place in the context of declining average prices. As reported in Table 8-2, the average price of gasoline sold in California declined substantially during the decade. From an average realized price of $1.17 per gallon in 1980, it fell to 86 cents per gallon in 1988,[3] a decline of over 25% in 8 years. Moreover, these figures are in nominal dollars, so the decline in real prices (corrected for inflation) would be even greater. Those lower prices, however, did not reflect fewer gallons sold of gasoline.

Further data on the average size of service stations in California is reported in Table 8-3. While that table shows a continuing decline in the number of stations throughout the 1980s, the numbers given by the U.S. Census are much smaller than the number of station permits reported for the same year. Many of the permits may have gone unused, or used in establishments whose major economic activity was something else, so that the Bureau of the Census allocated the establishment to a different industrial classification. The ones presented here are those for whom traditional service station activities comprise their principal business.

Within this group, the average number of employees increased by nearly 30% during these years; and also the number of stations with ten or more employees grew by 88%. While there were still relatively few firms with 50 or 100

---

[3] These figures are computed by dividing sales in dollars by the number of gallons sold.

TABLE 8–1   Service Station Sales in California.

| | | — Gasoline Sales — | | |
Year	Total Sales $ (millions)	Gallons (millions)	$ (millions)	Other Sales $ (millions)
1980	15,934	11,228	13,173	2761
1981	16,261	11,130	14,334	1927
1982	13,886	10,970	13,387	499
1983	13,423	11,146	12,612	811
1984	13,931	11,518	12,961	1319
1985	14,163	11,739	13,178	985
1986	11,613	12,272	10,962	651
1987	12,308	12,675	11,083	1225
1988	13,133	13,019	11,471	1662

SOURCE: California State Board of Equalization.

TABLE 8–2   Average Gasoline Sales per Service Station in California.[a]

| | ——— Sales ——— | |
Year	Gallons (thousands)	$ (thousands)
1980	711	834
1981	713	918
1982	719	878
1983	768	869
1984	810	912
1985	855	960
1986	916	819
1987	986	862
1988	1,097	942

[a] These averages are gasoline sales from Table 8-1 divided by the number of service station permits issued in California.

employees, a much larger number than before have more than ten employees. The decline in the number of service stations followed from an increase in the size of successful stations and not from any diminution in demand.

During the past decade, the number of service stations continued to decline even while the market share of independent marketers remained stable. As a result, there is no indication that such firms were responsible for the structural changes that occurred. However, there is evidence that repair facilities at service stations declined substantially. Such stations are indicated by those with

TABLE 8-3   Gasoline Service Stations in California.

Year	Total No. of Establishments	Avg. No. of Employees per Establishment	No. of Establishments with No. of Employees Exceeding:		
			10	50	100
1980	10,605	5.7	1245	33	6
1981	10,338	6.2	1439	24	7
1982	10,404	6.2	1494	29	7
1983	10,793	5.7	1494	26	4
1984	10,150	6.2	1713	29	4
1985	9,490	7.0	2018	43	6
1986	9,406	6.9	2079	38	8
1987	9,745	7.4	2346	53	10

SOURCE: County Business Patterns, California (various editions).

TABLE 8–4   Full-Service and Self-Service Distribution Margins.[a]

		Year of Highest Margin		
	Period	Full Service	Self Service	Correlation Coefficient
Los Angeles	1980–89	1989	1980	-0.70
San Francisco	1980–89	1989	1980	-0.84
San Diego	1980–89	1989	1980	-0.49
Bakersfield	1982–89	1989	1982	-0.57
Sacramento	1982–89	1989	1988	0.09
Stockton	1983–89	1989	1983	-0.57
Fresno	1983–89	1989	1983	-0.57
All Areas	1980–89	1989	1980	-0.52

[a] These distribution margins are average values for regular unleaded gasoline.

"zero service bays"; these stations accounted for 31% of the total number in January 1981 but 45% in July 1986 (Lundberg Surveys).[4] Many stations are now affiliated with convenience retail stores rather than automobile repair shops. Their primary economic activity is the sale of gasoline rather than automobile repair, and larger stations are a result of these revised attentions.[5]

---

[4]   These data refer to the entire United States.

FULL-SERVICE AND SELF-SERVICE DISTRIBUTION MARGINS

Distributions margins on the sale of gasoline are measured by the difference between retail and wholesale prices, after taxes are deducted. During the 1980s these margins varied greatly among different grades of gasoline and between full-service and self-service sales. In this section, we review the available evidence on changes in distribution margins within California during the decade.

By the end of the decade, in December 1989, regular unleaded gasoline accounted for 55% of all U.S. refinery sales.[6] We therefore first examine distribution margins on that grade of gasoline. Table 8-4 describes the pattern of distribution margins for regular unleaded gasoline in seven California cities as well as for all cities combined. As can be seen, there are striking differences in the trend of full-service and self-service margins. Full-service margins increased everywhere throughout the 1980s, while except for Sacramento, self-service margins declined everywhere.

This issue is explored further in Table 8-5 where actual margins throughout the state are reported. As indicated there, average distribution margins for full-service and self-service sales were not much different in the start of the decade. Full-service distribution provided only 10 to 15% higher margins than self-service distribution. By the end of the decade, however, distribution margins between these two types of sales were wildly different. The differences were on the order of 300 to 500% for unleaded gasoline and more than 1200% for regular leaded gasoline. While self-service margins had declined sharply for regular grades of gasoline, full-service margins had increased by over 170% in one case and over 200% in others.

These changes in distribution margins accompanied a major shift in consumer demands. Between 1975 and 1989, self-service sales increased steadily, from 22% to 80% of the total (National Petroleum News Factbook, 1990, p. 132). Furthermore, gasoline price controls ended with the expiration of the

---

[5] The decline in automobile repair facilities at service stations is not reflected in the recent trend in Automobile Repair Shops. The number of such establishments in California was 9,917 in 1980 but 14,699 in 1987. Furthermore, the number of employees in these establishments was 51,675 in 1980 and 67,661 in 1987. See U.S. Bureau of the Census (1980, 1987). Thus, there is no indication that the overall demand for automobile repair services in California declined during the 1980s. Instead, repair services were provided more frequently in specific facilities and less in service stations.

[6] The Lundberg Surveys report the following proportions of total sales: regular unleaded, 55.3%; regular leaded, 6.2%; middle unleaded, 10.0%; and premium unleaded, 28.5%.

TABLE 8–5   Average Distribution Margins in California. (cents per gallon)

| | Regular Unleaded Gasoline | | Premium Unleaded Gasoline | |
	Full Service	Self Service	Full Service	Self Service
January 1980	16.19	14.40	NA[a]	NA
January 1981	15.14	10.30	14.99	11.43
January 1982	20.76	8.64	19.72	13.01
January 1983	29.59	7.62	31.19	15.37
January 1984	30.47	7.10	33.38	15.67
January 1985	35.86	8.10	39.45	18.12
January 1986	36.64	7.07	39.96	15.05
January 1987	46.51	4.53	47.21	9.77
January 1988	44.06	6.65	43.81	10.53
January 1989	50.32	7.90	48.76	11.96

| | Regular Leaded Gasoline | | Premium Leaded Gasoline | |
	Full Service	Self Service	Full Service	Self Service
January 1980	15.83	13.74	16.10	14.45
January 1981	14.06	7.11	14.95	11.35
January 1982	18.40	5.77	19.45	15.07
January 1983	26.69	3.19	26.25	16.14
January 1984	27.94	2.72	24.33	14.21
January 1985	33.37	3.75	28.24	16.58
January 1986	33.35	2.87	28.67	14.34
January 1987	39.98	0.13	NA	NA
January 1988	37.79	2.90	NA	NA
January 1989	43.04	3.22	NA	NA

[a] NA indicates data are not available.

Emergency Petroleum Allocation Act in October 1981. As expected, prices rose with the end of price controls so these increases occurred in the 1980s.

To some extent, the rise in full-service margins resulted from higher labor costs. As reported by the Bureau of Labor Statistics, average wage rates in California cities for both "material handling laborers" and "motor vehicle mechanics" increased substantially over the decade. For the first occupation, the median increase in nominal wages exceeded 100% in Fresno between 1980 and 1989 but only by 37% in San Francisco and 87% in Los Angeles. For the second occupation, the median increase varied between 31% and 66% across six

California cities (U.S. Bureau of Labor Statistics). While substantial, in no case did average wage rates increase by the 200% change recorded for full-service gasoline distribution margins. Higher wage payments therefore are only part of the explanation.

Another factor is the shift in automobile repair services from service stations to repair shops. As cars became both more reliable and more sophisticated, stations provided fewer of these services. As a result, there are fewer repair employees available to provide full-service gasoline sales. Employees must be hired just for that job, which is far more costly.

This argument rests on the presumption of economies of integration between the supply of automobile repair services and the full-service distribution of gasoline. In this case, when service station repair services decline, the cost of full-service distribution increases; and so does the distribution margin required to cover such costs.

The changing pattern of full-service and self-service distribution margins may also be due to a new pattern of price discrimination. While the evidence is inconclusive, it is likely that the 12% difference in margins at the start of the decade was not sufficient to cover the added costs of full-service, so this difference indicated price discrimination in favor of full-service sales. By the end of the decade, however, pricing strategies had changed under the pressure of increased competition for self-service sales. As a result, distribution margins for these sales had declined by nearly 50%. At the same time, the remaining demand for full-service sales was sufficiently inelastic so that prices could be increased substantially.[7]

From the viewpoint of consumers, the changes that occurred in average distribution margins had a mixed effect. While those consumers content to use self-service outlets gained from the increased competition that occurred, consumers who sought additional consumer services were twice disadvantaged. Not only did the number of outlets decline sharply so there were fewer stations to choose from but also they paid much higher prices. The distribution sector provided less service with the gasoline that it sold and charged higher prices for those who required assistance.

---

[7] For two recent studies that find price discrimination in retail markets for gasoline, see Shepard (1991) and Borenstein (1991).

DISTRIBUTION MARGINS AND RETAIL PRICES ACROSS FIRMS

A critical issue for any distribution system is the manner by which wholesale prices affect retail prices. In this section, we explore this question empirically: first, through regression equations that measure the effect of wholesale prices of gasoline on retail distribution margins, and second, through equations that directly explain retail prices. In all equations, dummy variables are included for the seven major oil companies, with independent marketers omitted, and for six cities in California, omitting Stockton.

The first set of equations is reported in Table 8-6 where the dependent variable is the retail distribution margin for 10 years, for the seven major oil companies individually and the independent marketers together, and for seven cities in California. Not all years are available for each city, and not all companies sold each type of gasoline, so the number of observations is fewer than the theoretical maximum. Particularly in the case of premium leaded gasoline, there are many fewer observations as some companies did not distribute that product widely.

As reported in Table 8-6, there are eight estimated regression equations. A clear difference between the equations estimated for self-service sales and those for full-service sales is that the trend coefficient is uniformly negative in the former case but uniformly positive in the latter. These estimates confirm what we observed before, which is that distribution margins throughout the 1980s declined for self-service sales but increased sharply for full-service sales.

Of greater interest are the estimated coefficients that measure the effect of the station's buying price on the retail margin. As reported, these coefficients are negative in seven cases out of eight, and always significantly negative in that the estimated coefficients exceed twice their standard errors. Only in the case of self-service sales for premium unleaded gasoline is the coefficient positive, but it is not significant. The negative coefficients indicate that distribution margins decline when wholesale prices rise and increase when wholesale prices fall. For regular unleaded gasoline, the most common grade, an increase in the wholesale price of, say, 10 cents per gallon would lead to a decline in the retail margin of 1 cent per gallon on self-service sales and of 1.3 cents per gallon on full-service sales.

TABLE 8–6   The Determinants of Retail Margins, 1980–1989.[a]

	Unleaded Gasoline		Leaded Gasoline	
	Regular	Premium	Regular	Premium
Self-Service Sales				
Intercept	22.19	9.25	24.15	19.16
	(1.47)	(2.96)	(1.68)	(2.41)
Wholesale price	-0.099	0.030	-0.145	-0.058
	(0.013)	(0.023)	(0.015)	(0.023)
Trend	-1.192	-0.009	-1.671	-0.254
	(0.084)	(0.176)	(0.097)	(0.141)
R^2	0.43	0.19	0.48	0.14
N	538	415	469	182
Full-Service Sales				
Intercept	20.61	33.59	18.10	26.92
	(2.04)	(4.50)	(2.49)	(2.56)
Wholesale price	-0.132	-0.209	-0.096	-0.146
	(0.028)	(0.035)	(0.023)	(0.024)
Trend	2.931	2.293	2.318	2.915
	(0.116)	(0.266)	(0.143)	(0.151)
R^2	0.85	0.76	0.73	0.82
N	522	380	449	182

[a] The figures in parentheses are standard deviations of the coefficients. Each equation also includes six dummy variables to indicate the six major oil companies in California, the Independents being the seventh category, and also six dummy variables to indicate a particular city.

Another way of examining a similar effect is through the equations reported in Table 8-7. There, the dependent variable is the retail price rather than the distribution margin, while the explanatory variables are the same as before. The corresponding issue is whether the estimated coefficient on the wholesale price variable is greater or less than unity. As can be seen, this coefficient in the same seven equations as before is significantly below unity, although the significance level is problematical in the case of regular leaded gasoline on full-service sales. Again, the only aberrant estimate is for premium unleaded gasoline, for which the coefficient exceeds unity.

TABLE 8–7   The Determinants of Retail Prices, 1980–1989.[a]

	Unleaded Gasoline		Leaded Gasoline	
	Regular	Premium	Regular	Premium
Self-Service Sales				
Intercept	33.06	21.57	36.46	32.98
	(1.59)	(3.80)	(1.76)	(3.10)
Wholesale price	0.963	1.082	0.901	0.972
	(0.014)	(0.030)	(0.016)	(0.029)
Trend	-0.138	0.981	-0.711	0.944
	(0.091)	(0.226)	(0.101)	(0.182)
R^2	0.97	0.94	0.96	0.88
N	538	415	469	182
Full-Service Sales				
Intercept	35.49	47.57	30.20	41.12
	(5.92)	(5.26)	(2.81)	(2.88)
Wholesale price	0.889	0.827	0.951	0.879
	(0.051)	(0.041)	(0.026)	(0.027)
Trend	4.111	3.426	3.534	4.296
	(0.337)	(0.311)	(0.162)	(0.170)
R^2	0.42	0.71	0.80	0.92
N	522	380	449	182

[a] See footnote to Table 8-6.

For the leading grade of gasoline—regular unleaded gasoline—the estimated coefficients indicate that an increase in the wholesale price of 10 cents per gallon leads to an increase in the retail price of only 9.6 cents per gallon on self-service sales and only 8.9 cents per gallon on full-service sales. As before, these results indicate that the distribution sector absorbs a portion of any changes that occur in wholesale prices; retail prices, of course, follow wholesales prices, but any change in the former is less than that of the latter.[8]

---

[8]   This result appears more broadly in the industry. Between July 1986 and August 1987, crude oil prices increased by nearly 72% while wholesale gasoline prices increased by 37.5% and retail gasoline prices by only 11.8%. Between August 1987 and October 1988, crude oil prices declined by 33.1% while wholesale gasoline prices fell by 12.5% and retail gasoline

These results are consistent with the presence of either competitive or monopolistic markets. Even where stations have monopoly power, it is not generally profitable to raise their price as much as the higher wholesale price, nor to lower their price as much when wholesale prices decline. This fact is reflected in the empirical analysis. The findings reported here indicate that approximately 9 cents out of every 10-cent-per-gallon increase or decrease in the wholesale price in gasoline is passed on to consumers.

Tables 8-8 and 8-9 provide the estimated coefficients on the dummy variables also included in the regression equations that explain retail margins. In Table 8-8 are the estimated coefficients for each of the seven major oil companies where the omitted observation refers to the independent marketers reported as a group. As can be seen, the Arco margin is generally the lowest for self-service sales, with stations distributing Texaco and Union Oil products also showing relatively low distribution margins. The negative sign on the coefficients indicate that the average reported margins for these companies is less than average margin for the independents. While the distribution margins for the remaining oil companies are generally higher than the independents, not so for these three firms.

These results also indicate that although independent marketers may have lower retail prices than the service stations who distribute gasoline produced by the major oil companies, their distribution margins are not always lower when other factors are taken into account. For self-service sales, the independent marketers on average have higher margins than those of stations who distribute Arco, Texaco, and Union Oil products but lower margins than those of the four remaining companies. On full-service sales, the reported pattern in the coefficients is different. While Arco's margins are generally lower than the others except for Union Oil, they are typically higher than those reported for the independent marketers for most grades of gasoline.

Table 8-9 reports the estimated coefficients for six of the seven California cities included in the analysis. Stockton is omitted so that the equations could be estimated, and one can posit a zero coefficient (with the price effect equal to the intercept) in this case. As can be seen, for both full-service and self-service

---

prices by 3.9%. Finally, between October 1988 and March 1989, crude oil prices increased again by 52.2% while wholesale gasoline prices rose by 20.9% and retail gasoline prices by 4.0%. See Petroleum Marketers Association of America (1989).

TABLE 8–8   The Effects of the Major Suppliers on Distribution Margins.[a]

	Unleaded Gasoline		Leaded Gasoline	
	Regular	Premium	Regular	Premium
Self-Service Sales				
Arco	-1.99 (0.56)	-1.83 (0.70)	-1.61 (0.59)	-1.28 (1.05)
Chevron	2.04 (0.57)	2.37 (0.70)	0.98 (0.60)	-0.40 (1.13)
Exxon	1.62 (0.60)	1.93 (0.77)	1.01 (0.63)	0.10 (1.02)
Mobil	1.02 (0.57)	1.30 (0.71)	0.61 (0.60)	-0.33 (1.05)
Shell	0.63 (0.57)	0.13 (0.71)	-1.24 (0.60)	1.46 (1.05)
Texaco	-0.48 (0.59)	-1.39 (0.78)	-0.62 (0.62)	-1.38 (0.90)
Union	-2.08 (0.58)	-0.16 (1.00)	b	-0.74 (0.67)
Full-Service Sales				
Arco	2.28 (0.83)	0.81 (1.19)	1.82 (0.94)	-1.98 (1.09)
Chevron	7.27 (0.78)	5.92 (1.09)	6.07 (0.88)	-1.28 (1.12)
Exxon	5.02 (0.82)	4.46 (1.18)	4.61 (0.93)	-1.09 (1.10)
Mobil	5.61 (0.78)	4.54 (1.11)	4.90 (0.89)	-0.87 (1.16)
Shell	8.26 (0.20)	7.39 (1.11)	4.24 (0.89)	-1.17 (1.12)
Texaco	4.38 (0.81)	1.30 (1.20)	3.23 (0.92)	-1.35 (0.96)
Union	1.13 (0.79)	-0.68 (1.49)	b	-1.89 (0.73)

[a] Figues in parentheses are standard deviations of the coefficients.
[b] In these data, Union Oil made no sales of regular leaded gasoline.

TABLE 8-9   The Effects of Location on Distribution Margins.[a]

	Unleaded Gasoline		Leaded Gasoline	
	Regular	Premium	Regular	Premium
Self-Service Sales				
Bakersfield	1.19 (0.55)	2.01 (0.72)	0.28 (0.62)	0.82 (0.90)
Fresno	0.68 (0.55)	1.00 (0.72)	1.11 (0.62)	0.62 (0.90)
Los Angeles	-1.82 (0.54)	0.23 (0.70)	-0.73 (0.61)	0.26 (0.91)
Sacramento	-0.06 (0.54)	0.57 (0.70)	0.46 (0.61)	1.05 (0.91)
San Diego	0.14 (0.54)	2.67 (0.70)	1.48 (0.61)	1.33 (0.93)
San Francisco	0.70 (0.54)	2.03 (0.70)	1.11 (0.61)	2.39 (0.90)
Full-Service Sales				
Bakersfield	-0.54 (0.77)	-2.85 (1.12)	-3.85 (0.94)	0.19 (0.94)
Fresno	0.42 (0.77)	-0.13 (1.16)	1.17 (0.94)	0.02 (0.94)
Los Angeles	-2.21 (0.75)	-1.89 (1.06)	-3.00 (0.90)	-2.07 (0.96)
Sacramento	1.18 (0.75)	1.64 (1.08)	0.71 (0.91)	-0.79 (0.99)
San Diego	-1.22 (0.76)	-0.46 (1.09)	-1.50 (0.92)	-1.53 (0.98)
San Francisco	-0.28 (0.75)	0.68 (1.06)	-1.21 (0.91)	-0.15 (0.95)

[a] The figures in parentheses are standard deviations of the coefficients.

sales, the coefficient for Los Angeles is the lowest in the sample. Distributions margins in that city are substantially lower than those found elsewhere in California.

To examine this issue further, we determined the number of service stations per capita for these seven metropolitan areas as well as for California as a whole. Although Los Angeles had far more stations than any other city, it stood sixth out of seven in the number of stations per capita in both 1982 and 1987. And it had fewer stations per capita than California as a whole in both years. Moreover, Los Angeles had an intermediate number of stations in 1987 per registered vehicle: a larger number than Sacramento or San Diego but fewer than the San Francisco–Oakland metropolitan area.[9] The lower margins reported for Los Angeles are therefore not due to the presence of more stations per person or per vehicle.

To explore this issue further, we considered the demand side of the equation: whether cars are driven more so that gasoline purchases are higher in Los Angeles than elsewhere in California. Here again, the findings are unexceptional. The average car in Los Angeles county is driven 14,273 miles per annum as compared to the state-wide average of 14,434 miles. And the Los Angeles figure is slightly lower than that reported for Sacramento and San Diego although higher than that for the San Francisco–Oakland metropolitan area (California Department of Automotive Repair, 1991). Furthermore, the number of gallons of gasoline sold in the Los Angeles area either per registered vehicle or per person is actually lower than that reported for the other three areas (Petroleum Marketers Association of America, 1989). The primary difference between Los Angeles and the other cities is its much larger size. There is doubt, however, that this factor alone could explain the lower distribution margins found in that city.

THE EFFECT OF WHOLESALE PRICES

In the discussion above, we considered the impact of wholesale prices on retail prices and thereby on distribution margins. In this section, we extend the anal-

---

[9]   The number of establishments that sold gasoline are from the Census of Retail Trade (1982, 1987). The population data are published in the California Statistical Abstract. The number of registered vehicles as of December 31, 1987 is reported by the California Department of Motor Vehicles (1991).

ysis by examining the impact of wholesale prices on reported differences between full-service and self-service sales for the same grade of gasoline. For these sales, input prices are the same, since suppliers charge the same price regardless of the level of services provided with the product. These empirical results continue the prior analysis of full-service and self-service distribution margins, although here we deal directly with observed differences between the two prices.

The empirical findings are presented in Table 8-10. As reported, the coefficients on the trend variable are always positive and highly significant, which describes the same results found above: that the gap between full-service and self-service prices has grown sharply over the decade of the 1980s. In addition, we have significantly negative coefficients for three grades of gasoline out of four, with a significant positive coefficient for regular leaded gasoline. Where the reported coefficients are negative, that result indicates that the difference between full-service and self-service prices decline when wholesale prices increase but increase when these prices decline. Apparently, wholesale prices are more closely linked to self-service retail prices, so that these price differences diminish when wholesale prices increase but expand when wholesale prices fall. Surprisingly, the opposite result is found for a regular leaded gasoline, but this grade accounts for a small share of total sales.[10]

Insofar as the individual companies are concerned, we observe that the coefficients are often significantly positive for unleaded and regular leaded gasoline. This finding indicates that differences between full-service and self-service retail prices for the major oil companies are typically larger than those found for the independent oil companies, which are the basis of comparison in these equations. However, these results do not apply to premium leaded gasolines. For most sales, the price difference between full-service and self-service sales by the major brands is larger than that of the independent oil companies.

Finally, we consider the determinants of wholesale prices, which, of course, represent the major input into the distribution sector. These prices are affected by conditions at previous stages of production, as discussed in earlier chapters. However, we can still examine locational differences for such prices.

---

[10] See footnote 5 above.

TABLE 8-10   The Determinants of the Differences between Full-service and
             Self-service Prices.[a]

| | Unleaded Gasoline | | Leaded Gasoline | |
	Regular	Premium	Regular	Premium
Intercept	-1.48	25.42	-6.58	8.49
	(2.43)	(4.48)	(2.30)	(2.03)
Wholesale Price	-0.040	-0.253	0.052	-0.095
	(0.021)	(0.035)	(0.027)	(0.019)
Trend	4.40	2.45	4.27	3.41
	(0.14)	(0.26)	(0.17)	(0.12)
Arco	5.04	2.59	3.75	-0.75
	(0.99)	(1.19)	(1.13)	(0.88)
Chevron	5.64	4.04	5.33	-0.89
	(0.93)	(1.08)	(1.06)	(0.95)
Exxon	3.73	3.08	3.78	-1.43
	(0.98)	(1.19)	(1.12)	(0.86)
Mobil	5.07	3.65	4.66	-0.77
	(0.94)	(1.11)	(1.07)	(0.91)
Shell	8.29	8.03	5.76	0.14
	(0.93)	(1.10)	(1.07)	(0.88)
Texaco	5.17	3.09	3.99	-0.16
	(0.97)	(1.20)	(1.11)	(0.75)
Union	3.52	-0.30	NA[b]	-1.54
	(0.94)	(1.49)	0.0	(0.75)
R^2	0.86	0.79	0.78	0.90
N	520	378	449	179

[a] The figures in parentheses are standard deviations of the coefficients. Each equation also includes six dummy variables to indicate a particular city.
[b] In these data, Union Oil made no sales of regular leaded gasoline.

We estimated regression equations for each of the four grades of gasoline to explain the variation in wholesale prices. The independent variables included both a time trend and dummy variables to represent each of the seven cities mentioned above. As expected, the coefficient on the trend variable is always significantly negative, which indicates only that prices declined throughout the decade.

Of greater interest are the coefficients for six cities in California, with Stockton the omitted variable. The estimated coefficients are generally smaller than their standard errors and never statistically significant. In effect, these equations demonstrate that wholesale prices are generally the same throughout California. Thus, the lower prices found in Los Angeles and the lower distribution margins reported there, are not associated with lower wholesale prices. The lower prices paid by consumers in Los Angeles are due entirely to conditions in the distribution sector.

SOME FUTURE CONCERNS

As reported elsewhere in this volume, the question of alternative fuels that can be used to power automobiles is frequently raised. And an important aspect of this concern is whether the current distribution system would be a serious impediment. This matter is relevant because of the special needs of automobile drivers for a widespread network of fuel supply points. They would not accept alternative fuels unless a widespread distribution system were in place.

Much would depend on the auspices under which the new fuel was supplied. If the product was supported by one or a number of the major oil companies, then the lessee dealers who are associated with them would surely introduce the new fuel as rapidly and as easily as they did when alternative grades of gasoline were introduced. The cost of pumping a liquid fuel from an underground tank to a customer's automobile should be little different.

Furthermore, the relevant distribution margins should be similar. These margins are not set on a percentage basis but rather represent an absolute amount per gallon sold of gasoline, although they tend to be higher on self-service sales of premium rather than regular grades. Still, there is little reason to expect that the margins associated with the distribution of alternative fuels would be much different than that observed for gasoline.

To the extent, however, that an alternative fuel is offered by a different entity than a major oil company, the results could be different. Many dealers are effectively tied to their supplier because of limited space. In such cases, they would be less likely to sell the products of rival suppliers. Moreover, stations supplied and often owned by independent marketers are more closely tied to their affiliated suppliers. It might therefore be difficult to achieve rapid distribution of an alternative fuel if the supplying oil companies had no economic interest in doing so.

A second concern regarding the distribution system concerns its response to external shocks. Major shocks occurred during the 1970s as various crude oil-supplying nations restricted exports for a substantial period of time. As a result, supplies were limited and prices rose sharply. Similarly, in March of 1989, the *Exxon Valdez* ran aground in Valdez harbor and spilled ten million gallons of crude oil into Prince William Sound. In both cases, retail prices jumped sharply and there were charges of price gouging levied at station owners.

Whatever the validity of such charges elsewhere in the industry, they are not supported by the available evidence on the distribution sector. As reported above, service stations ameliorate rather than magnify changes in the wholesales price of gasoline. The distribution system has softened the effects of past oil shocks on consumers.

In the 1980s, a different kind of shock appeared—a continued decline in average gasoline prices throughout the decade—yet the distribution system remained on course. Substantial changes did occur during the decade (the number of stations declined significantly and average station size increased; full-service margins rose sharply while self-service margins declined), but there was no indication that these changes resulted from the decade-long decline in retail prices.

While these various episodes led to cries for public intervention into the market place, there is little reason to respond to such calls. By all accounts, the distribution system performed as it was designed to do. Particularly important was the fact that it ameliorated the changes in wholesale prices that occurred. At the same time, we have little explanation of why distribution margins were significantly lower in Los Angeles and whether these low margins could be achieved elsewhere in California. If so, there is the prospect for still lower distribution costs. Although the retail sector works well, there is still opportunity for improvement.

REFERENCES

Borenstein, S. (1991). "Selling Costs and Switching Costs: Explaining Retail Gasoline Margins." *Rand Journal of Economics*, Vol. 22, pp. 354–369.

California Department of Automotive Repair (1991). Telephone conversation.

California Department of Motor Vehicles (1991). Telephone conversation.

California State Board of Equalization. "Taxable Sales in California," various editions. Sacramento.

California Statistical Abstract (various editions). Sacramento: Economic Development Agency of the State of California.

Census of Retail Trade (1982). U.S. Department of Commerce, Bureau of the Census. Washington, D.C.: Government Printing Office.

Census of Retail Trade (1987). U.S. Department of Commerce, Bureau of the Census. Washington, D.C.: Government Printing Office.

County Business Patterns, California (1980). U.S. Department of Commerce, Bureau of the Census. Washington, D.C.: Government Printing Office.

County Business Patterns, California (1987). U.S. Department of Commerce, Bureau of the Census. Washington, D.C.: Government Printing Office.

Lundberg Surveys.

Miller, R. A. (1963) "Exclusive Dealing in the Petroleum Industry: the Refiner-Lessee Dealer Relationship." In *Yale Economic Essays*, Vol. 3, pp. 223–247.

National Petroleum News (1990). Vol. 82, No. 7. New York: McGraw-Hill.

Petroleum Marketer's Association of America (1989). "Recent Increases in the Prices of Petroleum Products, Exhibit VI, Significant Price Increases in Petroleum Products Since the Valdez Oilspill." Hearings before the Subcommittee on Energy Regulation and Conservation, Committee on Energy and Natural Resources, United States Senate, April 17, pp. 175–186. Washington, D.C.: Government Printing Office.

Shepard, A. (1991). "Price Discrimination and Retail Configuration." *Journal of Political Economy*, Vol. 99, No. 1, pp. 30–53.

U.S. Bureau of Labor Statistics (selected issues). "Area Wage Surveys." Washington, D.C.: Government Printing Office.

U.S. Department of Energy (1981). "The State of Competition in Gasoline Marketing." Washington, D.C.: Government Printing Office.

INDEX

A

STUDIES IN INDUSTRIAL ORGANIZATION

Series Editors:
H.W. de Jong, *University of Amsterdam, The Netherlands*
W.G. Shepherd, *University of Massachusetts, Amherst, U.S.A.*

Publications:
1. H.W. de Jong (ed.): *The Structure of European Industry.*
   Revised edition 1988, see below under Volume 8
2. M. Fennema: *International Networks of Bank and Industry*
   (1970-1976) 1982   ISBN: 90-247-2620-4
3. P. Bianchi: *Public and Private Control in Mass Product Industry.*
   The Cement Industrial Cases. 1982   ISBN: 90-247-2603-4
4. W. Kingston: *The Political Economy of Innovation.*
   (1984) 1989 2nd printing   ISBN: 90-247-2621-2
5. J. Pelkmans: *Market Integration in the European Community.*
   1984   ISBN Hb:90-247-2978-5;
                                          ISBN Pb:90-247-2988-2
6. H.W. de Jong and W.G. Shepherd (eds): *Mainstreams in Industrial*
   Organization. 1986
   Book I: *Theory and International Aspects*
                                          ISBN: 90-247-3461-4
   Book II: *Policies, Antitrust, Deregulation and Industrial.*
                                          ISBN: 90-247-3462-2
                        Set ISBN Book I + II: 90-247-3363-4
7. S. Faltas: *Arms, Markets and Armament Policy.* The Changing Structure of
   Naval Industries in Western Europe. 1986   ISBN: 90-247-3406-1
8. H.W. de Jong: *The Structure of European Industry.* 2nd revised ed.
   (of Volume 1).1988   ISBN Hb:90-247-3689-7;
                                          ISBN Pb:90-247-3690-0
9. I.L.O. Schmidt and J.B. Rittaler: *A Critical Evaluation of the Chicago*
   School of Antitrust Analysis. 1989   ISBN: 90-247-3792-3
10. B. Carlsson (ed.): *Industrial Dynamics. Technological, Organizational*
    and Structural Changes in Industries and Firms. 1989
                                          ISBN: 0-7923-9044-X
11. Z.J. Acs and D.B. Audretsch (eds): *The Economics of Small Firms.* A
    European Challenge. 1990   ISBN: 0-7923-0484-5
12. W.J. Kingston: *Innovation, Creativity and Law.* 1990
                                          ISBN: 0-7923-0348-2
13. B. Dankbaar, J. Groenewegen and H. Schenk (eds.): Perspectives in
    Industrial Organization. 1990   ISBN: 0-7923-0814-X
14. P. deWolf (ed.): *Competition in Europe.* Essays in Honour of
    Henk W. de Jong. 1991   ISBN: 0-7923-1050-0
15. C. van der Linde: *Dynamic International Oil Markets.*
    Oil Market Developments and Structure 1860-1990. 1991
                                          ISBN: 0-7923-1478-6

16. D.B. Audretsch and J.J. Siegfried: *Empiricial Studies in Industrial Organization*. Essays in Honor of Leonard W. Weiss. 1993
                                                    ISBN: 0-7923-1806-4
17. R. J. Gilbert (ed.): *The Environment of Oil*. 1993
                                                    ISBN: 0-7923-9287-6